Advanced Agroecology

Songliang Wang • Claude D. Caldwell

Advanced Agroecology

Springer

Songliang Wang
Fujian Agriculture and Forestry University
Fuzhou, Fujian, China

Claude D. Caldwell
Dalhousie University
Valley, NS, Canada

ISBN 978-3-032-08688-4 ISBN 978-3-032-08689-1 (eBook)
https://doi.org/10.1007/978-3-032-08689-1

This Springer imprint is published by the registered company Springer Nature Switzerland AG
The registered company address is: Gewerbestrasse 11, 6330 Cham, Switzerland

This book is dedicated to the students of the FAFU-Dalhousie University/NSAC 2 + 2 BSc Joint Program since the grade 2003–2024, whose enthusiastic and active participations of agroecology course and program and the success after graduation both in Canada and China motivated and enlightened the two authors' devotion to this book.

Preface: The Fiat Emperor Needs a New Cloth!

In the Hans Christian Andersen (1837) fairy tale, the untainted child told the popular phenomenon in society that 100 times spoken lies would not change their facts called 'The Fiat Emperor Has No Clothes'. For modern humankind who has not heard the story, here is the gist recalled by Yoder-Wise P.S. (2022).

> A wealthy emperor is approached by two charlatans who convince him that they are master weavers. Their message is that their special clothes are invisible to those who are not brilliant. Naturally, the emperor and other key allies consistently 'see' the fabric as it is being woven and then when the emperor is fitted in his new suit. The emperor sets off on a processional through the city wearing his invisible new clothing, while his assistants carry the imaginary train. As the emperor marches through the city, all of the townspeople remain silent for fear of appearing stupid—that is, until a child yells out that the emperor has no clothes.

The tale could be true in today's economic sectors and academies. It is quite evident that economic data characterized predominantly by gross domestic production (GDP) means little and yet we have been fooled by the preaching of economists. In March of 2008, Scientific American published the short article 'The Economist Has No Clothes', Robert Nadeau, the author, wrote in a hard-hitting tone:

> The theory (the neoclassical economics) created in the 19th now serves as the basis for coordinating activities in the global market system, was developed by adapting equations from 19th-century physics that eventually became obsolete. Unfortunately, it is clear that neoclassical economics has also become outdated. The theory is based on unscientific assumptions that are hindering the implementation of viable economic solutions for global warming and other menacing environmental problems.

The unscientific assumptions of the neoclassical economics include at least but not limited as follows:

> ① The market system is a closed circular flow between production and consumption, with no inlets or outlets. ② Natural resources exist in a domain that is separate and distinct from a closed market system, and the economic value of these resources can be determined only by the dynamics that operate within this system. ③ The costs of damage to the external natural environment by economic activities must be treated as costs that lie outside the closed market system or as costs that cannot be included in the pricing mechanisms that operate within the system. ④ The external resources of nature are largely inexhaustible, and those

that are not can be replaced by other resources or by technologies that minimize the use of the exhaustible resources or that rely on other resources. ⑤ There are no biophysical limits to the growth of market systems.

The long-term neglect of natural capital and environmental parameters in the neoclassical economics inevitably leads to environmental problems and crises. Among them, global climate change is the biggest and most crucial one which has been the effect of human development. When those crises became unprecedentedly devastating, the successors of neoclassical economics stand out as the Saviours to put forward a serious so-called reform and transformation to reverse the situation in terms of old financial routes, such as carbon trading and GHGs emission trading. But as great Albert Einstein once said (Caldwell and Wang, 2020, preface, vi):

We cannot solve our problems with the same thinking as used to create them.

Most people in this world believe Albert's golden saying, like the Fiat Emperor, the economist has no clothes; unfortunately, nowadays we seem to be lacking the untainted child to tell the truth. The world system runs ahead on the same pathway.

It is particularly true in the agriculture and food system arena, under the name of feeding the increasing population, enhancing the food yield and production has become the unique and lofty commitment of generations to generations by producing and applying high technologies. Obviously, it has not solved the problem; according to FAO (2021), there were still 820 million people lacking sufficient daily food, 925 million consume less than 1800 calories per day, inevitably leading to 9.9% of total population in 2020 compared to 8.8% in 2019 suffering from chronical undernourishment, 3 billion people were affected by micronutrient deficiencies, and food price is still mounting since the financial crisis in 2008 and worsens since COVID-19. The commitment of humans to increase crop yield and food production has not changed the facts that humankind is still in jeopardy in terms of food security both in quantity and quality.

Maybe it is good for our humankind to ask the question 'is agriculture a good idea?' The famous Ecologist, Jared Diamond raised the question in 1987, he states critically that the question might be stupid in the view of the progressivists, and reasonable in the view of ecologists, but now evidential in the view of the archaeologists. Research shows that one of our hunting and gathering ancestors ate 75 plants and animals as his or her daily food with far more balance in nutrition than modern people who consume high carbohydrate and starchy food, such as rice and potato cultivated by agriculture. Why are we so certain 10,000 years of agriculture is much better than 2 million years' hunting and gathering life choice? How do we know the lives of people 10,000 years ago had better life when they gave up hunting and gathering? Furthermore, archaeologists affirmed the modern gross social and sexual inequality, the disease, and despotism coming with agriculture based on their archaeological materials (Diamond 1987).

Therefore, it is time to invite the untainted children to tell the truth; the current version of Agroecology textbook will take the role. The intended audience of *Advanced Agroecology* is postgraduate students in universities and colleges, particularly in agriculture and food-associated universities and colleges after they follow and read the authors' *Introduction to Agroecology* published by Springer in 2020. This book also can act as the reference for those consumers who recognize their right and power in shaping a sustainable food system.

The text is divided into 10 chapters:

Chapter 1 *connects agroecology to the major problems of world agriculture and food system.*

Chapter 2, 'The "What", "How," and "Why" of Agriculture and Agroecology', *narrates the current patterns and consequences of world agriculture and agroecology's role in finding a way out.*

Chapter 3, 'Agroecosystems and Their Services', *addresses that agroecosystem services which are using ecology methodology to capitalize the natural value can be a leverage for finding a sweet spot between human health, environment justice, and economy in agriculture and food system sector.*

Chapter 4, 'Agricultural Biodiversity and Food Security', *infers that agrobiodiversity is a key in the fields for food security when human consider the latter in a broad view.*

Chapter 5, 'Agroecology and Agri-Food System Transformation', *elaborates the agroecology's role in global and regional agriculture and food system transformation towards their sustainable development.*

Chapter 6, 'Integrating Rural Landscape Sustainability Into Agroecology', *incorporates agroecology from field landscape to rural landscape to reach the rural community vibrant.*

Chapter 7, 'New Challenges for Education in Agroecology, *reviews agroecology education in global and continental context.*

Chapter 8, 'Systematic and Participatory Action Research in Agroecology', *reinvents the research orientation in agriculture and food system under the school of agroecology.*

Chapter 9, 'Agroecology for Food Sovereignty: A Science, A Practice, and A Social Movement', *updates agroecology to multiple dimensions in response to systematic problems of world agri-food system.*

Chapter 10, 'Eco-agriculture: Adapting to Global Climate Change by Applying Agroecology , *concludes with eco-agriculture standing for the praxis component of agroecology when it was coined as applying ecology to find an alternative model substituting fossil-based modern one, but this time needs to confront to the urgent climate change.*

All in above, let's welcome the more untainted children moving to the foreground to call forth the collective action if humankind truly want a sustainable future.

Fuzhou, Fujian, China Songliang Wang
Valley, NS, Canada Claude D. Caldwell

Acknowledgements

As this book is about to be published, we would like to express our gratitude to a great number of people who were involved with and offered help for our agroecology co-teaching adventure, among them, two elegant ladies are on our prior list, Ms. Shannon L. Kilyannek, a Senior Instructor of former Nova Scotia Agricultural College, who was consistently professional, hardworking, and uncomplaining, instructing laboratory class for agroecology course, and Ms. Haiyan Su, FAFU-NSAC joint program coordinator, who was always making smooth and considerate schedules for our international agroecology teaching team every time each year. Definitely, we would like to thank our nearly one and half thousand students who were recruited to the joint program, because of them, we have been called as 'the teacher', the respectful title in China. Thanks also go to the Springer editors and coordinators, Mr. Aaron Schiller, Ms. Nivetha Moorthi, Mr. Henry Rodgers, and Mr. John Barthelemy, and the anonymous reviewers of this textbook, without their effort, this book will be far from exquisite.

Special thanks must be given to our spouses, RongJi Qiu and Lynn Caldwell. Their support, patience, and love have given us the strength to try to make a difference through communication and teaching.

The production of this textbook was supported by a grant on Building a Full English Postgraduate Course in Agroecology [grant number: 712018270336], funded by Fujian Agriculture and Forestry University in 2020.

Competing Interests The authors have no competing interests to declare that are relevant to the content of this manuscript.

Contents

About the Authors

Songliang Wang is a Professor at Fujian Agriculture and Forestry University (FAFU) in China. He has been teaching agroecology to undergraduates since 1990 and initiated an Advanced Agroecology course in 2013 for Chinese and international postgraduates majoring in agriculture and agronomy. Since 2004, Dr. Wang has been a teaching partner with Dr. Claude Caldwell for the Sino-Canadian 2 + 2 educational program. Together, they have published three agroecology textbooks in English since 2012 and more than ten papers related to agroecology teaching and research since 2009. Dr. Wang is an editorial board member of the *International Journal of Agricultural Sustainability*. In 2019, he initiated the grassroot professor plan to expand his classroom to rural fields, which brings university students who were learning agroecology course to the field to co-learn agroecology and eco-agriculture with new farmers to fuel and facilitate agroecological transformation in Fujian province. Under his efforts, a group of agroecology experts were trained and several county scale eco-agriculture bases were established in Fujian province.

Claude D. Caldwell is a Professor Emeritus from Dalhousie University, who initiated an agricultural ecosystem course for all freshmen undergraduates in Nova Scotia Agricultural College in the early 1990s. He conceived a new understanding of agriculture in 1996, i.e. Agriculture is the science, art, politics, and sociology of changing sunlight into happy, healthy people, which closely mirrors todays' agriculture and food system transition. Dr. Caldwell was a co-founder of Sino-Canadian 2 + 2 educational program in 2003 and co-taught agroecology for the programmed students with Dr. Wang from 2004 to 2024. He has been active in international development work since 1995 and worked in 11 different countries over the past 28 years He has taught and/or developed undergraduate and graduate curriculum in cooperation with universities in Jamaica, Trinidad, The Philippines, Indonesia, Vietnam, The Gambia, Tanzania, Cambodia, Ethiopia, Uganda, and China. He continues to be active in teaching and research with international colleagues, especially in China.

Chapter 1
Agroecology and the New Four Horsemen of the Apocalypse

Abstract Humanity faces four interwoven critical challenges: the need to double food availability by 2050 sustainably; global climate change; decreasing quality and quantity of freshwater; and economic livelihood, all of which form an existential threat to society. Transformations in food and agriculture systems are key solutions. This chapter will explore the interconnected ways that agroecology can inform and lead international strategies and local actions to execute a second real Green Revolution, decrease greenhouse gases (GHGs) emissions, and stimulate local and global economies.

Keywords The Anthropocene · The New Four Horsemen · Agri-food system · Elephants in the room · Climate change

Learning Objectives

After reading this chapter, students should be able to:

1. Identify and explain the 'elephants' hidden in the Anthropocene room.
2. Recognize the dangers and opportunities inherent in our present Anthropocene era and identify positive ways to make our civilization sustainable for centuries to come.
3. Reflect on the importance of 'The New Four Horsemen' in terms of food security and sovereignty under the school of agroecology.

****

> The best way to predict the future is to create it.
> —Peter Drucker

> If we really give a damn, we must grasp the nettle.
> —Colin Tudge 2014

In the twenty-first century, humanity stands at a crossroads, confronted with four interwoven critical challenges that pose an existential threat to society. By 2050, the

S. Wang, C. D. Caldwell, *Advanced Agroecology*,
https://doi.org/10.1007/978-3-032-08689-1_1

global population is projected to reach nearly 10 billion, necessitating a sustainable doubling of food availability. However, the currently prevailing agricultural practices, which rely heavily on fossil-based inputs, are exacerbating global climate change. Agriculture contributes approximately 25% of global greenhouse gas (GHG) emissions, if plus with food consumption sector, the number reaches to more than 33%, with deforestation for farmland expansion and excessive use of fertilizers being major culprits. Simultaneously, the quality and quantity of freshwater are in decline. Over two billion people already lack access to safely managed drinking water, and agricultural irrigation accounts for about 70% of global freshwater withdrawals, often leading to water scarcity and pollution. Economically, rural livelihoods are under threat, with small-scale farmers struggling to compete in global markets dominated by industrial agriculture.

Transformations in food and agriculture systems (hereafter refers to 'agri-food system') emerge as key solutions to these intertwined crises. This chapter delves into the pivotal role of agroecology, an approach that integrates ecological principles into food and farming systems. Agroecology offers a holistic framework to inform and lead both international strategies and local actions, aiming to initiate a second truly 'Green' Revolution. By adopting agroecological practices or ecological agriculture, we can decrease GHG emissions, enhance soil health, and conserve water resources. Moreover, agroecology has the potential to stimulate local and global economies by promoting value-added agricultural products, creating green jobs, and strengthening local food systems. In doing so, we firstly need to have a cognitive transformation on agriculture and food system.

1 Recognizing 'Elephants in the Room'

Over the last 10,000 years, agriculture has been a key positive force for health, development of society, culture, and happiness across the globe. Unfortunately, within the past 100 years, agriculture has also become a significant source of many of humanity's problems. Agriculture is therefore also a large part of the solution to those problems; however, we need new thinking and actions using all the tools of technology and all the lessons of the past woven together. However, over the past 2 decades, the following changes have occurred:

- **World Population Has Increased from Six Billion to Eight Billion**: However, despite the ambitious goals to eliminate poverty and hunger presented in the UN Millennium Development Goals (MDGs) in 2000 and subsequently in the UN Sustainable Development Goals (SDGs) of 2015, progress has been slow in confronting increasing environmental and social problems (Sachs et al. 2009; Sachs 2012; Xu et al. 2020). In the year 2000, approximately 28% of the global population was living in extreme poverty; this decreased by the year 2024 to around 9%. However, in 2024, the wealthiest 1% of the world's population controls 48% of global household wealth, up from 40% in the year 2000. Inequality is a global pressure (Barrett et al. 2022). Meanwhile, in the year 2000, food insecurity affected approximately 10% of the global population; by 2024, this rose to an estimated 19% (HLPE 2020; FAO 2021, 2023).

- **The Concentration of Carbon Dioxide in the Atmosphere Rose from 370 to 427 ppm:** Carbon dioxide is one of the key greenhouse gases (GHGs) responsible for climate change (Den Elzen and Schaeffer 2002; Fuhrer 2003). According to the report of the IPCC (Intergovernmental Panel on Climate Change), the average temperature in 2000 was 0.42 °C above the twentieth century average vs 2024 at 1.5 °C above preindustrial levels (Intergovernmental Panel on Climate Change 2007, 2023). Climate change has heavily affected global vegetation, led to systematic changes of agroecosystems, and, as a result, produced changes to the food production system (Parmesan and Yohe 2003). Climate change therefore is a huge barrier to the realization of food security and the other Sustainable Development Goals.
- **Between 2000 and 2024, the Availability of Freshwater for People Has Decreased Dramatically**: In 2025, 5 billion people suffer from water stress. This fact alone will challenge world food security (Brown 2004).
- **From 2000 to 2019, There Were 7348 Major Natural Disasters Around the World:** By comparison, the previous 20-year period, 1980–1999, had 4212 natural disasters. Over the past 5 decades, human activities for economic growth, food and other daily material needs like wood and fibre have changed the natural ecosystem to a great extent (Millennium Ecosystem Assessment 2005), altering the balance of nutrient cycles and meteorological processes, resulting in more frequent extreme weather and natural disasters than what would be expected.

Global food insecurity has been exacerbated due to unfolding wars and rising costs of energy, fertilizers, and food. The world emerged from the COVID-19 pandemic and reawakened to climate change and the threat it poses to a return to 'normal' life. The overused term 'new normal' is treated with fear and apprehension. One of the keys to achieving a better 'normal' is to engage the world's best minds in agriculture. A positive future is possible, but successful strategies are needed to address these truly interwoven and interdependent challenges.

'Elephant in the room' is an English-language idiom for an obvious problem or risk that people want to avoid. We would like to use the image of elephants in the room to help define what changes are needed. Consider the following global issues which do not receive the amount of discussion and reflection needed to make such change:

- **Capital-Driven Growth:** Economic development and capitalism have been the predominant ideologies affecting social life and modern economics for many years. Every action is measured by monetary value, including agriculture and food systems. For example, the raising of a pig for food is valued uniquely only for the meat; therefore, the use of antibiotics, synthetic compounds, and potential food system contaminants is justified on the basis of monetary efficiency. Ecosystem services of a farming system are therefore lost in the evaluation of the over-simplified monetary process.

- **Production-Oriented Blinders:** Due to the increasing growth of world population, 'Feed the World' narratives have become the main target of agricultural production. In most developing countries, this narrative has become the political mission and the overriding raison d'être for the agricultural industry. With that approach in mind, food security was interpreted as simply the amount of food production (IPES-Food 2016). This has hampered the establishment of sustainable food systems since it is in contrast with holistic strategies like agroecology.
- **Technological Darwinism:** Technology is a double-edged sword and unfortunately, over the last several decades, we often have been blinded by the positive side of technology applications in favour of production increases. Many of those highly productive technologies have had their social costs and environmental costs, such as soil degradation, environmental pollution, and decreased quality of food.
- **Rampant Consumerism**: Classic growth-based economics proposes that, to sustain capital-driven growth, a luxury life is equated with a good life, leading to the popularization of hedonistic consumerism (Durning 1997). In the past, population increase was seen as an elephant in room; nowadays, per capita consumption is a more prominent elephant that no one wants to address effectively.
- **Freshwater Usage and Pollution:** Water is one of the most precious resources for human life; freshwater resources on earth are limited. Although 70% of our planet is covered with water, the drinkable and usable water is limited to less than 2% of total water in the global ecosystem (Wang and Caldwell 2020). Within that 2%, 99% is locked in the glacier of earth's north and south poles, only 1% water is available for human drinking and agricultural irrigation. Irrigation for agricultural purposes accounts for 75% of worldwide freshwater consumption. In this regard, the quantity of water resource has become an increasing issue; a significant portion of our underground water resource has been contaminated by human activities. Many of those detrimental activities are linked to agriculture, including runoff from fields which include pesticides and fertilizers. Considering the pressures on freshwater from human activities and the increasing demand for water by an increasingly water consuming population, water might be the biggest elephant in the room.
- **Peasant, Farmer, Agricultural Labourer or Farm Worker as a Series of Titles for Those Endeavouring to Produce Food for the Rest of Us:** In many parts of the world, the occupation of producing food is not valued highly. Whether they are called peasant, farmer, agricultural labour, or farm worker, often the intent is that these are lesser occupations. Many farm workers are exploited with poor living conditions and low wages and little or no protection under the law in many countries. In some areas of the world, child labour is still a very real part of the agricultural sector. Consequently, worldwide, many of these workers abandon the land and move to the city to sell their labour for a living wage that allows them to support their families and provide for their children's education so that they do not have to become farmers. The real question then becomes 'who will produce our food?' Worldwide, therefore, the livelihood of the farmer might

be considered another important elephant in our agri-food room. For sustainable agricultural food systems, those who produce food must be treated fairly and must have a life that is both financially and socially as important as what they produce.

- **A Few Giant Capital-Driven Companies or Many Small-Scale Farmers; Who Contributes More?** The global agri-food system is increasingly dominated by a few large corporations, to the overall financial detriment of consumer groups and society in general. The dominance of such agricultural oligarchies has, in many countries, eliminated small-scale farmers. National protectionism, supported by subsidies for those dominant companies produces unfair international trade policies. This results in destruction of sustainable agricultural systems and farmer livelihood. This continues now to spread into the environment of developing countries (Anbumozhi and Radhakrishnan 2005; Bacon 2005; Babin 2015). In this regard, economic system inequality is also an elephant in our agri-food room.
- **Animal Production and Animal Welfare**: In natural ecosystems, animals are indispensable trophic levels in cycling matter and facilitating the flow of energy to maintain the ecosystem. In agricultural systems, animal production has played the role in a similar way, but its primary role is to produce meat, milk, and eggs for human consumption. Under the philosophy of luxury life equals good life, large-scale animal production systems are encouraged to expand throughout the world taking up most of the primary production base. This initiative has had a large impact on food security and environment (Hedenus et al. 2014; Charles et al. 2018); it has led to food being produced cheaply but often very inhumanely, leading to significant ethical concerns (Jensen and Sørensen 1999; Thompson and Nardone 1999). Once again this is something we do not want to discuss but has large impact on agriculture and people.

All these 'elephants' are interlinked. The common thread is the dominance of production systems which value profit over principle. Although each one is challenging, it is important to acknowledge the problems and discuss what needs to happen to create more sustainable and equitable agri-food systems.

2 Harnessing 'Horsemen of the Apocalypse'

'Horsemen of the Apocalypse' is an ancient image from the Christian Bible meant to alert people of the dangers of wrong paths being taken. They are often interpreted as harbingers of the end times, each representing different forms of disaster that will befall the Earth. Today we are met with four new intertwined Horsemen of the Apocalypse:

- Increasing demand for food due to population increase and consumerism.
- Weather chaos due to climate change.

- Decreasing quality and quantity of freshwater.
- Human livelihood disruption.

All these issues are significantly negatively affected by our current global agriculture model. Hence, a sustainable agricultural transformation and food system restructuring strategy is the catalyst for success in all four areas. Our food system can be transformed through a knowledge-based agenda.

To find a positive path forward, the New Four Horsemen of the Apocalypse need to be met with the concepts of agroecology, which may be defined as the science of our time (Wang et al. 2012, preface). In general, the consistent, successful model in responding to these challenges follows principles of agroecology broadly defined as a science, a practice, and a social movement (Wezel et al. 2009, see Chap. 9 in detail). It is comprised of some key characteristics:

1. **Change from Isolated Disciplinary Thinking to Systematic Thinking:** This should begin in elementary schools with universal eco-civilization education instilling educated respect for nature, people, food, and the relationship of people with food (Francis et al. 2001; Wang and Wu 2022). It needs to be solidified and expanded in tertiary institutions by linking social sciences and natural sciences integrating into ecology and social change theory and practice (Qiu et al. 2023).
2. **Integration of Methods from Natural Sciences and Social Sciences, Building a Resilient Holistic Food System**: The relative isolation of the disciplines from each other in today's tertiary institutions has inadvertently produced a knowledge and expertise gap between natural and social sciences (Wezel et al. 2014). This divide produces a barrier to solving real-world interdisciplinary problems and questions. This interconnectivity of problems and questions is particularly evident in food systems, in which the human side (farmer and family) interacts closely with the natural basis of all food value chains (field and soil).
3. **Shift from a Theoretical Paradigm to One Which Emphasizes Field Teaching and Learning Processes**: Education and the application of the principles we have learned and tested in agroecology will be key to our global response to the interrelated changes observed over the past 2 decades. Only through education, careful planning, and global cooperation, we can find positive answers to the questions that we face. The increase in community supported agriculture (discussed in Sect. 3 of this chapter) in China since 2008 was initiated by well-educated youths, who are cordially called 'the new farmers' (because they are different from old generation farmers and doing 'the "new" agriculture'). They usually are a group of the reawakened consumers who realized that the current food system was on the wrong path, gaining the motivation to learn agroecology and eco-agriculture. These students have worked with innovative leaders from college or university faculties to bring the educational resources to agricultural fields and rural areas (Wang 2023).

3 Shifting Paradigm for Agriculture

This will mean educating ourselves, society at large, and the next generation, to look at food systems through the three lenses of sustainability (environmental, economic, and societal), livelihood, and equity (Fairness).

To illustrate the new paradigm for agriculture, consider the following three examples drawn from the global South. These demonstrate careful thought, innovation, and sensitivity to the needs of the community that have produced positive results. The first (*solar-based irrigation systems*) highlights a simple case of the need for the adaptation of appropriate new technology to decrease fossil fuel consumption, increase efficiency, and stabilize production (DeMarco 2018). The second (*community-based integrated pest management, CBIPM*) looks at the critical role of community in any changes in agronomic techniques for more sustainability (Caldwell and Kilyanek 2002). The third (*community supported agriculture, CSA*) demonstrates how community, production techniques, education, and value chains link to produce sustainable long-term agroecological systems (China Environment Net 2022; Filho et al. 2020).

3.1 Solar-Powered Irrigation in Africa (with a Focus on Smallholders)

In many parts of Africa, the agriculture sector contributes about 30% of gross domestic product (GDP) and accounts for 80% of national employment, mainly in rural areas (Bellwood-Howard and Ripoll 2020). In such situations, GDP growth originating from agriculture is considerably more effective in reducing poverty compared with GDP originating from outside agriculture. Small-scale farming is largely rain-fed and thus highly vulnerable to climate change impacts such as unreliable rainfall and frequent episodes of drought. Solutions at this individual small grower level have the potential for maximum impact on communities.

Solar-powered irrigation is an opportunity to reduce dependency on fossil fuels, improve efficiency in irrigation at lower costs, and stabilize production (Hartung and Pluschke 2018).

A next-generation technology approach has been tested in Senegal, Burkina Faso, and Cameroon and involves affordable, appropriate technology at the small farm level (DeMarco 2018). Solar panels have become cheap and accessible; however, commercial, specially made small-scale solar pumps have remained expensive. DeMarco's approach uses off-the-shelf DC pumps that can work directly with solar panels so that there is no need for a battery or controller. This makes the system simple, portable, and less susceptible to breakdown. Figure 1.1 shows a 250 W system operating in Cameroon; however, smaller systems, 60 W, have been shown to operate well at a lower cost.

Fig. 1.1 250 W system operating in Cameroon (DeMarco 2018, used with permission)

3.2 Community-Based IPM (CBIPM) in Vietnam

For a small landowner to implement an integrated pest management (IPM) program, it is essential to have communication and cooperation from their neighbours. Figure 1.2 shows a community consultation regarding IPM methods in Vietnam. The three basic elements of CBIPM are illustrated in Fig. 1.3. Most CBIPM activities involve at least two of these elements: training/learning processes, organizing/group development, and experimentation/knowledge generation, although different elements are prominent in different activities. Together, the three elements of CBIPM lead to farmer empowerment (Mahlein 2016). To use the words of one IPM farmer-trainer from Indonesia, this type of IPM helps farmers to 'stand on their own and think for themselves to do their own field observations, make their own discoveries, make their own decisions, and take action on their own'.

Examples of community IPM activities where the *training/learning process* is prominent are:

- Farmer field schools (FFS) conducted by IPM farmers for other farmers.
- Rural school curriculum that incorporates IPM concepts.
- Functional literacy programs that include IPM.

Examples of activities where *experimentation/knowledge generation* is prominent are:

- Insect zoos and compensation studies, managed by farmers as part of FFS organized by the government or non-governmental organizations (NGOs).
- Field studies that are organized and implemented by groups of FFS graduates.

Fig. 1.2 Community consultation regarding IPM methods in Vietnam (Caldwell and Kilyanek 2002)

- Action research facilities involving several studies carried out by IPM farmers over a number of cropping seasons.

Examples of activities where *organizing/group development* is prominent are:

- IPM farmer clubs, associations, and congresses.
- Planning and technical meetings organized by farmers.
- Farmers' advocacy and efforts to mobilize funding from local government in support of community action.

Community IPM started among Asian rice farmers who wanted to solve pest management problems, but it has since developed in several directions. Other crops, such as vegetables, maize, soybean, cotton, and tea, have also become the focus of training and experimentation managed by IPM farmers. Issues such as soil fertility, water management, and marketing have sometimes become just as important as solving pest problems, if not more so. Finally, it is not just in Asia where community IPM is happening. As a result of international exchanges by IPM trainers and program managers, farmer field schools and community-based IPM programs have started in several African countries and in parts of the Near East. More recently, urban community IPM has become reality. These new initiatives involve the integration of appropriate level technologies such as pest prediction software that can be shared on cell phones.

Participatory evaluations in Vietnam of the long-term impact in nine community-based IPM districts showed that the farmers had reduced their pesticide use by 65% and economized on seeds. Generally, the farmers also obtained the same or better

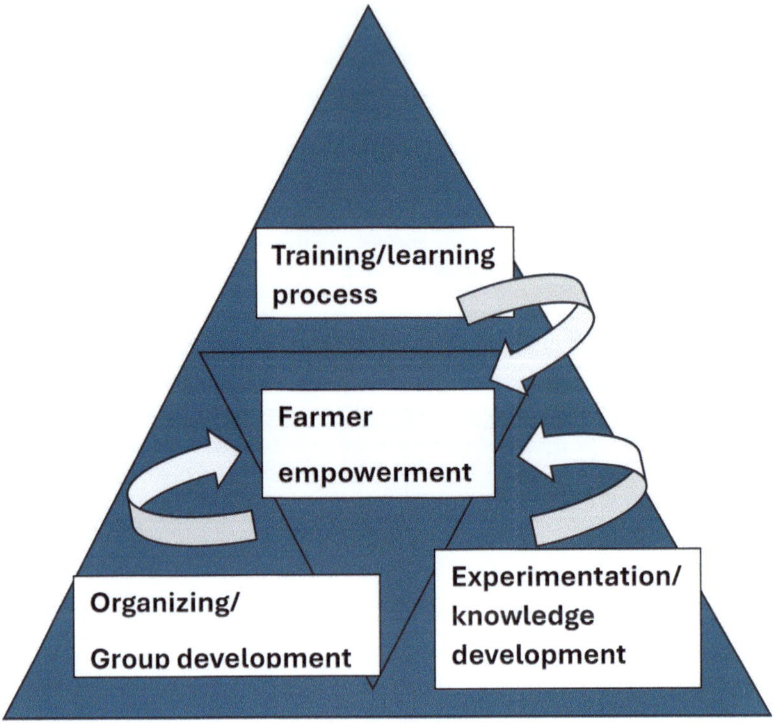

Fig. 1.3 The three basic elements of community-based integrated pest management (CBIPM) (from Caldwell and Kilyanek 2002)

yields than non-IPM farmers, with net income per hectare up 24% (Quyen et al. 2002).

A revolution in the way rural people think about their local environment is crucial. An understanding of native ecological and biological systems is pertinent for this transformation of rural thinking. This knowledge provides rural communities with a basis for making management decisions and performing activities that are sustainable and productive. A recent evaluation of CBIPM in fruit production in Vietnam emphasizes the need to link these practices with the overall value chain as the next part of the evolution of sustainable production (Truc et al. 2021).

3.3 Community Supported Agriculture (CSA) in China

In response to population pressure, food security is of paramount concern for China. The adoption of Green Revolution techniques from the mid-twentieth century, while increasing production, has also caused widespread environmental damage and food safety issues. At the same time, producers and consumers have become both physically and psychologically distant from each other due to the impositions of middlemen who often occupy an unfair portion of the value chain. One case study in Fujian province of China showed that farmers in the province share just 5% of the average total value of the agroecosystem services they created. In addition, the food carbon footprint has been increasing dramatically due to the twin effects of agrochemical inputs and long-distance food miles.

One way of closing the gaps in our food system is an approach called community supported agriculture (CSA). It is a way of directly connecting producers and consumers by allowing the consumer to subscribe to the harvest of a certain farm or group of farms. The producer and consumer share both the risks and the benefits of farming. The producer provides a weekly or biweekly box of products from the farm in return for the consumers' subscriptions. The amount of subscription varies according to the CSA model. This is not just an economic association but one that fosters communication and trust.

Over the past decade, schools of agroecology in China have stimulated and supported flourishing CSA communities. Starting in 2008, multiple food chain issues were further ignited by the notorious event of melamine in commercial milk powder and the tragic poisoning of many young Chinese. This was the wake-up call for rebuilding food safety, fair agricultural value chains, and lowered food carbon footprint. The first CSA in China was initiated in a Beijing suburb by a doctoral candidate (Ms. Shi Yan) majoring in agricultural economics who was learning and practising agroecology in the USA. She initiated this at her own expense and named it 'The Little Donkey' (Shi et al. 2011). The authors are proud that the next successful CSA was established in a suburb of Fuzhou by Mr. Wei Chang, a Fujian Agriculture and Forestry University (FAFU) graduate majoring in agronomy. He attended our introduction to agroecology course in his second college year and then organized a university level organic farming learning and studying society in his third college year (Wang et al. 2019). This CSA, named Jiamei Family Farm has grown from one family farm to encompass several adjacent farms which supply the produce collectively for the CSA. The families cooperate in the production and marketing of produce but also are active in education of consumers through field days, consumer visits, and targeted hands-on educational sessions on the farm.

Since then, CSA models have flourished nationwide in China. Recently, more than 800 CSA initiatives have sprung to life around China, thanks to the work of a growing network of young, new farmers (China Environment Net 2022). Consequently, these CSA models have acquired the following positive impacts in China's agroecosystem and food value chain:

1. **Return to Mutual Trust Between Producer and Consumer**: In the Wei Chang CSA family farm 30 kilometres from the centre of Fuzhou city, at the time when he started to run the farm at 2009, there were only ten consumers from the city community who prepaid their vegetable quota. Gradually, the number grew to 75 by 2010 (Wang et al. 2019), Even the industrialized style famers were transitioning their chemical-based farming to ecological compatible farming after having mutual evening learning sessions with Wei Chang.

2. **Increased, Stabilized Income for Producers and Fair Price for Consumers**: In the CSA model, unlike the industrial and marketing model, all the money paid by the consumer went to those new farmers' pocket; this has helped greatly to solve the economic problems in rural area (Wang 2019).
3. **Improved Agroecosystem Services**: The CSA model encourages the new farmer financially to do good agricultural practice, and in return, agroecosystem services, such as increased biodiversity in the field and improved soil quality (Villavicencio-Valdez et al. 2023).
4. **Decreased Food Carbon Footprint.** CSA practices involve a decrease in the use of various fossil-based inputs to the food system. For example, the shortened value chain removes the middleman since the produce is sold directly to the consumer. This means that the distance food travels will be decreased and will save fuel used in the transportation and preservation process (Cederlof 2016). Many studies indicate that a rice-fish production system, which is frequently adopted in CSA models, has decreased methane emission greatly and made a significant contribution in greenhouse gas emission and carbon footprint mitigation (Lu and Li 2006; Moucheng et al. 2014).

These three simple examples illustrate the need for careful thought and education to new possibilities, i.e. thinking our way to a better future.

3.4 Transformation Strategy

Sustainable transformation in agriculture must prioritize the interactive good health of humans, animals, and the Earth. The major barriers to a 'new better normal' are gaps in shared knowledge and a lack of global cooperation. In a time when greater international consensus and action is needed, the game of zero-sum nationalistic politics threatens to scuttle effective action. The new normal must be a better normal led by the next generation of the brightest and best agriculturalists. Our food system can be transformed through a knowledge-based agenda via a new(second) truly 'Green' Revolution in agriculture that responds to the requirement for doubling of available food by 2050.

The (first) Green Revolution of the mid-twentieth century was historic in its global approach to facing widespread famine in many countries of the world. Politicians and scientists at that time uniformly understood food security to be the total amount of food in the world. With that focused approach, scientists broke through the yield constraints of crop varieties in maize, wheat, and rice to breed the dwarf and strong hybrid combinations (Dahlberg 1979). However, what those scientists and politicians did not anticipate was the incidental costs to society and the environment. The Green Revolution of the late 1950s and 1960s was extremely successful in transferring technology into the global South to increase food production at a critical time of mass famine. Genetic improvements of cereal crops, introduction of synthetic pesticides, increased irrigation, and use of inorganic fertilizers saved hundreds of millions of lives. While this was an amazing achievement, the Green Revolution is now often vilified, due to its unintended social, economic, and environmental impacts. The scientists and politicians at the time did not foresee that the production packages with high-energy use, intensive irrigation, increased chemical use, and land consolidation would greatly shift the balance of power in many nations of the global South. In addition, fertilizers and pesticides increased health issues and water contamination. This first Green Revolution went wrong because the well-meaning initiatives transferred systems that, in the long term, were not sustainable, not economically viable and did not meet society's need for safe and nutritious food within vibrant communities. The result was degradation of natural resources, the environment, and community. When we think of addressing our present food crises, a new Green Revolution naturally comes to mind but we need to remind ourselves of the complexity of the social, economic, and environmental systems and learn from our previous mistakes before we take actions (Horlings and Marsden 2011).

A climate-smart food production strategy increases available food while lowering GHGs emissions. Since agriculture is a significant contributor to global greenhouse gas (GHG) emissions, any new food production strategy needs to both respond and mitigate the unprecedented global warming. The term climate-smart agriculture was proposed to respond to the challenge (Zhang 2024); before this recognized uniform term, low-carbon agriculture was popular as a term with a similar connotation and objective as climate-smart agriculture (Wang et al. 2010). Under the general strategy of climate-smart agriculture, we now use such terms as carbon-capture agriculture (Zhang 2024) and climate-smart soil management (Lal 2004). However, the principal and difficult goal of carbon neutrality of agri-food system remains. In general, there are two branches of climate-smart agriculture, one is mitigation (Lal 2004) and another is adaptation (Debray et al. 2018).

A water-adapted agricultural model produces food with a particular emphasis on the efficiency of water use and reverses the degradation of our freshwater resource. Availability of clean freshwater is one of the keys to eco-civilization society building. In 2013, the UN proposed the concept of water security, which was defined by the United Nations as (2013):

> The capacity of a population to safeguard sustainable access to adequate quantities of acceptable quality water for sustaining livelihood, human well-being, and social economic development, for ensuring protection against water borne pollution and water related disasters, and for preserving a climate of peace and political stability.

To produce enough food for an increasing population, we use an increasing amount of water for irrigation on an expanding agrarian land base. A significant 75% of human freshwater is used for irrigation. There are two challenges linked to water usage in irrigation: water use efficiency and water contamination. It is essential to reverse this trend by improving both water use efficiency and improving the quality of watercourses. This approach is called 'smart' irrigation.

Irrigation is one of the key reasons we have been successful in feeding 7+ billion souls. While only 20% of the world's farmland is irrigated, it produces 40% of our food supply (Wang and Caldwell 2020). The highest yields obtained from irrigation are more than double the highest yields from rain-fed farming. Even low-input irrigation is more productive than high-input rain-fed farming. We need to continue to use irrigation to produce food, but this overuse trend needs to be reversed. Engineers and irrigation technologists will need to develop techniques and adapt present technology to irrigate more land, not less, but do it smarter, with less water use. That water may be from nonconventional sources such as harvested rainwater, recycled wastewater, and desalinized waters. From the technology standpoint, improved on-farm water management means having accurate measures of soil water and plant stress monitoring. It will involve the precision application of water and will need to be combined with appropriate agronomic practices, such as mulching and cover crops. Traditional husbandry methods have a lot to offer in terms of water conservation. This will take good communication between farmers and technologists, both in sharing information and in developing workable solutions.

3.4.1 An Ecology-Oriented Value Chain Development That Enhances Productivity, Recognizes Stakeholder Participation and Builds Equity and Food Sovereignty

The movement of agricultural products including the addition of processing, packaging, transportation, and marketing is called a value chain, since value is added to the products at each step in the process. Obviously, it is not truly a chain but rather a network that has a group of stakeholders with vested interests in the success of that network. The network includes producers, processors, wholesalers, retailers, consumers, and regulators. To understand a value chain, one must understand 'value' beyond just the monetary measure. Agriculture and the resultant web of interactions have impacts beyond economics, and value needs to be measured in terms of environmental and social impacts as well.

Any agricultural system that is not fair is not sustainable. If anyone in the value chain is not rewarded fairly for their investment in money and labour, the system is not sustainable in the long term. Those of us in agriculture are committed to making our food systems sustainable. We put great emphasis on environmental and economic sustainability, but we sometimes miss this idea of fairness. Fairness and equity in the food value chains remain a core concept of agroecology. Agroecology, in considering any food system, uses a value chain management (VCM) approach. A supply chain is a passive description of flow of material, whereas a value chain is a more active economic approach to understanding a sequence of events, including understanding the role and impact of research and development, design, and approaches to marketing. VCM establishes channels of communication among the stakeholders to ensure that the chain or web is one that is efficient and sustainable. Without good communication, each participant in the value chain tends to take a competitive approach and not a cooperative one. Fairness in the chain is one major indicator of sustainability. Collaboration among the various stakeholders along the food value chain is more important than ever for reasons of food safety, efficiency, especially energy, reducing losses of food, waste management, and food price control. Rural communities are dependent on just such a fair and equitable value chain. Value chains are better conceived to be intricate value webs.

All those in the value chain, especially consumers have a responsibility to at least halve food loss wastage. If we are to meet our goals of feeding our population in the future, it is essential to substantially decrease what is being lost. This will involve both technology and public policies in local, national, and international food value chains. In countries where much of the food waste is at the restaurant, food service, or home level, the onus is on the consumer to take the responsibility to decrease waste. Reducing the consumption of animal products can also improve the ability of the food system to supply the growing population. The application of basic concepts of agroecology will help develop the plan but the execution of the plan will rely on people throughout the value chain stepping forward to make it work.

The role of mid-stream intermediaries (such as processors, wholesalers, and logistics providers) is crucial in the structural transformation and economic development of fair, sustainable value chains. The changes in value chains have mixed impacts on smallholder farmers, with some benefiting from new market opportunities, while others face challenges due to increased competition and standards. Therefore, policy interventions should focus on improving infrastructure, market access, and regulatory frameworks to support inclusive growth in agri-food value chains. Agri-food value chain transformations are complex and multifaceted; this emphasizes the need for targeted policies to ensure equitable benefits across different stakeholders (Barrett et al. 2022).

3.4.2 A Value Chain Which Builds on the Concept of Food Sovereignty

Agroecology and food sovereignty are two key components in the new Green Revolution, proposed by Holt-Giménez and Altieri (2013, see also in detail in Chap. 9 of this book). The term food sovereignty was first used in 1996 by *La Via Campesina,* a transnational movement of small-scale farmers, peasants, agricultural workers, and Indigenous groups that subsequently defined it as 'the right of peoples to healthy and culturally appropriate food produced through ecologically sound and sustainable methods, and their right to define their own food and agriculture system' (Walker et al. 2021). In short, food sovereignty is the right of people to have control over their own food systems, including the production, distribution, and consumption of food. It is based on local food economies, sustainable practices, and cultur ally appropriate foods. The priority is on empowerment of local producers and consumers to determine and develop food systems that support local and national markets. In terms of production methods, such food systems rely heavily on environmentally friendly and socially equitable sustainable practices. There is also an emphasis on the importance of food that is culturally appropriate, respecting the traditions and tastes of the community. Food sovereignty proponents work to advocate for policies that support small-scale farmers and fishers, rather than become part of a larger corporate-controlled food system.

Food sovereignty and food security are related concepts that differ in their approaches but are not mutually exclusive (Mitchell 2024). In fact, some would say that food security is the goal and food sovereignty is the method to achieve it. In terms of definition, food security involves ensuring that all people have access to sufficient, safe, and nutritious food to maintain a healthy life. It focuses on the availability, access, utilization, and stability of food supplies (Maudrie et al. 2023). Food sovereignty emphasizes the right of people to define their own food systems. It focuses on local control, sustainable practice, and culturally appropriate food. The priority is on the needs and rights of local producers and consumers in contrast to the global market demands. Most nations work to achieve food security through global trade and large agricultural infrastructure. This has been shown to lead to environmental destruction and exploitative conditions around the world. The food sovereignty movement therefore resists corporate control and trade-based system and instead advocates for local, sustainable food production using ecological methods and social justice approaches. Food security is primarily concerned with quantity and quality of food available, with little consideration of the production methods. In contrast, food sovereignty approaches stress the importance of culturally appropriate food and social conditions of the food production, including the rights of the producers and the local communities, valuing indigenous knowledge. With that in mind, it is obvious that most national food security policies focus on ensuring food availability by using subsidies, trade agreements, and the application of technology. Food sovereignty requires policies in support of local food systems and the rights of small producers with the aim of an equitable and sustainable local food. In short, the food sovereignty approach to food security is through a transformed food system that is fair and more sustainable, with local control.

However, one looks at developing sustainable food systems, at every stage of food value chains, and one needs to look through three lenses: the lens of sustainability; the lens of livelihood; and the lens of equity. It is our goal for a production system to show positive results through all of these three lenses.

4 Concluding Remarks

In the global South, a successful new Green Revolution will employ appropriate level technology. What is an appropriate level will be determined by the site-specific requirements. For small landowners, it may involve technologies like the solar-powered irrigation pumps and pest prediction software available on cell phones. Higher technology solutions linked to drones and site-specific evaluations of crop stress may be more applicable in other areas. Whatever technology is adapted will need to be done in a two-way communication with farmers and with appreciation of indigenous knowledge and experience. The example of CBIPM shows the power of community to solve environmental and economic problems collectively. The model of the CSA takes technology, communication, and value chain management to a level where a true Green Revolution is possible. This approach produces a fair and open value chain which is of major benefit to small landholders, the true core food producers of the global South.

The major barrier to the successful application of agroecology is not knowledge but a lack of local communication and global cooperation. Agriculture-associated universities and colleges are recognizing the need for pedagogical transformation to reflect the new reality and the need for graduates with strong cross-disciplinary knowledge and skills. Anyone looking for a challenge and opportunity to make a difference should consider a career in the many aspects of agriculture, both scientific and social. To make real change will involve re-evaluating our past behaviour, considering new pathways to sustainability, educating ourselves, and finding opportunities from crises.

References

Anbumozhi, V., and J. Radhakrishnan. 2005. Effects of agricultural free trade on environmental quality. *Outlook Agriculture* 34 (2): 91–96.

Babin, Nicholas. 2015. The coffee crisis, fair trade, and agroecological transformation: impacts on land-use change in cost a rica. *Agroecology and Sustainable Food Systems* 39 (1): 99–129. https://doi.org/10.1080/21683565.2014.960549.

Bacon, Christopher. 2005. Confronting the coffee crisis: can fair trade, organic, and specialty coffees reduce small-scale farmer vulnerability in Northern Nicaragua? *World Development* 33 (3): 497–511.

Barrett, Christopher B., Thomas Reardon, Johan Swinnen, and David Zilberman. 2022. Agri-food value chain revolutions in low- and middle-income countries. *Journal of Economic Literature* 60 (4): 1316–1377.

Bellwood-Howard, Imogen, and Santiago Ripoll. 2020. Divergent understandings of agroecology in the era of the African Green Revolution. *Outlook Agriculture* 49 (2): 103–110.

Brown, Lester R. 2004. *Outgrowing the earth: the food security challenge in an age of falling water tables and rising temperatures*. New York: W.W. Norton.

Caldwell, Claude D., and Shannon L. Kilyanek. 2002. *Community-based integrated pest management*. A report on the community-based conservation management project in China and Vietnam. Internal document of Nova Scotia Agricultural College.

Cederlof, Gustav. 2016. Low-carbon food supply: the ecological geography of Cuban urban agriculture and agroecological theory. *Agriculture and Human Values* 33:771–784.

Charles, H., J. Godfray, Paul Aveyard, Tara Garnett, Jim W. Hall, Timothy J. Key, Jamie Lorimer, Ray T. Pierrehumbert, Peter Scarborough, Marco Springmann, and Susan A. Jebb. 2018. Meat consumption, health, and the environment. *Science* 361:eaam5324.

China Environment Net. 2022. *Regenerating farming and community agriculture in China* [WWW Document]. https://china-environment-news.net/blog/2022/03/06/regenerating-farming-and-community-agriculture-in-china/#:~:text=InChinainthelast10-15years%2Cabasisofthismovement%2Cwhichissproutingrapidly. Accessed March 1, 2023.

Dahlberg, Kenneth A. 1979. *Beyond the green revolution: the ecology and politics of global agricultural development*. New York: Plenum Press.

Debray, Valentine, Alexander Wezel, Adeline Lambert-Derkimba, Katia Roesch, Geir Lieblein, and Charles Andrew Francis. 2018. Agroecological practices for climate change adaptation in semiarid and subhumid Africa. *Agroecology and Sustainable Food Systems*. https://doi.org/10.1080/21683565.2018.1509166.

DeMarco, J. 2018. Small-scale solar irrigation is now affordable. In *International Forum on Solar Technologies for Small-Scale Agriculture and Water Management*. Rome.

Den Elzen, Michel, and Michiel Schaeffer. 2002. Responsibility for past and future global warming: uncertainties in attributing anthropogenic climate change. *Climatic Change* 54:29–73.

Durning, Alan. 1997. *How much is enough: the consumer society and the future of the planet*. Changchun: Jilin People's Publishing House.

FAO. 2021. *The state of food security and nutrition in the world. Transforming food systems for food security, improved nutrition and affordable healthy diets for all*. Rome: UN Food and Agriculture Organization.

FAO. 2023. *The state of food security and nutrition in the world. Transforming food systems for food security, improved nutrition and affordable healthy diets for all*. Rome: UN Food and Agriculture Organization.

Filho, F. H. I., W. B. Heldens, Z. Kong, and E. S. De Lange. 2020. Drones: innovative technology for use in precision pest management. *Journal of Economic Entomology* 113:1–25. https://doi.org/10.1093/jee/toz268.

Francis, Charles A., G. Lieblein, J. Helenius, L. Salomonsson, H. Olsen, and J. Porter. 2001. Challenges in designing ecological agriculture education: a Nordic perspective on change. *American Journal of Alternative Agriculture* 16:89–95. https://doi.org/10.1017/s0889189300008985.

Fuhrer, Jürg. 2003. Agroecosystem responses to combinations of elevated CO_2, ozone, and global climate change. *Agriculture, Ecosystems and Environment* 97:1–20.

Hartung, H., and L. Pluschke. 2018. *The benefits and risks of solar powered irrigation—a global overview*. FAO report of Rome conference. Rome.

Hedenus, Fredrik, Stefan Wirsenius, and Daniel J. A. Johansson. 2014. The importance of reduced meat and dairy consumption for meeting stringent climate change targets. *Climatic Change* 124:79–91.

HLPE. 2020. *Food security and nutrition: building a global narrative towards 2030*. A report by the High-Level Panel of Experts on Food Security and Nutrition of the Committee on World Food Security, Rome.

Holt-Giménez, Eric, and Miguel A. Altieri. 2013. Agroecology, food sovereignty, and the new green revolution. *Agroecology and Sustainable Food Systems* 37:90–102. https://doi.org/10.1080/10440046.2012.716388.

Horlings, L. G., and T. K. Marsden. 2011. Towards the real green revolution? Exploring the conceptual dimensions of a new ecological modernization of agriculture that could 'feed the world'. *Global Environmental Change* 21:441–452.

Intergovernmental Panel on Climate Change (IPCC). 2007. Climate Change 2007: Synthesis Report. *Contribution of Working Groups I, II and III to the Fourth Assessment Report of the Intergovernmental Panel on Climate Change* (CoreWrit. Team, RK Pachauri, A Reisinger, eds.), 104. Geneva: IPCC.

Intergovernmental Panel on Climate Change (IPCC). 2023. Sections. In *Climate Change 2023: Synthesis Report. Contribution of Working Groups I, II and III to the Sixth Assessment Report of the Intergovernmental Panel on Climate Change* [Core Writing Team, H. Lee and J. Romero (Eds.)], 35–115. Geneva: IPCC. https://www.ipcc.ch/report/ar6/syr/downloads/report/IPCC_AR6_SYR_LongerReport.pdf. Accessed May 17, 2025.

IPES-Food. 2016. *From uniformity to diversity: a paradigm shift from industrial agriculture to diversified agroecological systems*. http://www.ipesfood.org/images/Reports/UniformityToDiversity_FullReport.pdf. Accessed December 12, 2024.

Jensen, Karsten Klint, and Jan Tind Sørensen. 1999. The idea of "Ethical Accounting" for a livestock farm. *Journal of Agricultural and Environmental Ethics* 11:85–100.

Lal, R. 2004. Soil carbon sequestration to mitigate climate change. *Geoderma* 123:1–22.

Lu, Jianbo, and Xia Li. 2006. Review of rice–fish-farming systems in China—One of the Globally Important Ingenious Agricultural Heritage Systems (GIAHS). *Aquaculture* 260:106–113.

Mahlein, A. K. 2016. Present and future trends in plant disease detection. *Plant Disease* 100:1–11.

Maudrie, Tara L., Cassandra J. Nguyen, Rachel E. Wilbur, Megan Mucioki, Kaylee R. Clyma, Gary L. Ferguson, and Valarie Blue Bird Jernigan. 2023. Food security and food sovereignty: the difference between surviving and thriving. *Health Promotion Practice* 24 (6): 1075–1079. https://doi.org/10.1177/15248399231190366.

Millennium Ecosystem Assessment (MEA). 2005. *Ecosystem and human well-being: synthesis*. Washington, DC: Island Press.

Mitchell, Farrelly. 2024. *Food sovereignty vs food security*. https://farrellymitchell.com/food-security-consultants/food-sovereignty-vs-food-security/. Accessed May 17, 2025.

Moucheng Liu, Yin Xiong, Zheng Yuan, Qingwen Min, Yehong Sun, Anthony M. Fuller.2014. Standards of Ecological Compensation for Traditional Ecoagriculture: Taking Rice-Fish System in Hani Terrace as an Example. *Journal of Mountain Science*,11(4): 1049–1059

Parmesan, Camille, and Gary Yohe. 2003. A globally coherent fingerprint of climate change impacts across natural systems. *Nature* 421:37–42.

Qiu, Rongji, Xiaobin Zhu, and Songliang Wang. 2023. Re-structuring curriculum in tertiary institutions to drive sustainable eco-civilization in China. *African and Asian Studies* 22:11–22.

Quyen, P.B., T. Yem, Lam P. Van, T.T. Phuong, T.T. Huyen, H.T. Mai, Y. Guang, and F. Chang. 2002. *Community based integrated pest management—final report for community-based conservation management Chinese-Vietnam-Canada Tier 1 development project*. Nova Skotia Agricultural College and Fujian Agricultural University Joint Program by CIDA, Canada. Internal Document.

Sachs, Jeffrey D. 2012. From millennium development goals to sustainable development goals. *Lancet* 379:2206–2211.

Sachs, Jeffrey D., Jonathan E. M. Baillie, William J. Sutherland, Paul R. Armsworth, Neville Ash, John Beddington, Tim M. Blackburn, Ben Collen, Kevin J. BarryGardiner, H. Charles Gaston, J. Godfray, Rhys E. Green, Paul H. Harvey, Brett House, Sandra Knapp, Noëlle F. Kümpel, David W. Macdonald, Georgina M. Mace, James Mallet, Adam Matthews, Robert M. May, Owen Petchey, Andy Purvis, Dilys Roe, Kamran Safi, Kerry Turner, Matt Walpole, Robert Watson, and Kate E. Jones. 2009. Biodiversity conservation and the millennium development goals. *Science* 325:1502–1503.

Shi, Yan, Cunwang Cheng, Peng Lei, Tiejun Wen, and Caroline Merrifield. 2011. Safe food, green food, good food: Chinese community supported agriculture and the rising middle class. *International Journal of Agricultural Sustainability* 9:551–558.

Thompson, P. B., and A. Nardone. 1999. Sustainable livestock production: methodological and ethical challenges. *Livestock Production Science* 61:111–119.

Truc, Ngo Thi Thanh, Đoan Tran Oanh Bao, and Đo Thi Hoai Giang. 2021. Farmers' behavior and intention to adopt integrated pest management in fruit: case study in Vietnam. *Journal of Applied Science and Advanced Engineering* 17 (11): 755–769.

United Nations. 2013. *What is water security? Infographic*. https://www.unwater.org/publications/what-water-security-infographic. Accessed December 20, 2024.

Villavicencio-Valdez, Gabriela Valeria, J. Jacobi, M. Schneider, M. A. Altieri, and H. Suzán-Azpiri. 2023. Urban agroecology enhances agrobiodiversity and resilient, biocultural food systems. The case of the semi-dryland and medium-sized Queretaro City, Mexico. *Frontiers in Sustainable Food Systems*. 7:1066428. https://doi.org/10.3389/fsufs.2023.1066428.

Walker, C., L. DeMatteis, and A. Lienert, eds. 2021. *Selecting value chains for sustainable food value chain development—guidelines*. Rome: FAO. https://doi.org/10.4060/cb7623en.

Wang, Songliang. 2019. Concerted development of ecological agriculture along with community-supported agriculture to facilitate rural vitalization. *Chinese Journal of Eco-Agriculture* 27 (2): 212–217.

Wang, Songliang. 2023. The demand for the agriculture associated universities and colleges' teaching &learning transformation and educational evaluation reformation under the rural revitalization strategy: exemplified by agroecology course teaching. *China Agricultural Education* 24 (3): 8–16.

Wang, Songliang, and Claude D. Caldwell. 2020. Water as the basis for cultivated ecosystems. In *Introduction to Agroecology*. pp. 49–60 https://doi.org/10.1007/978-981-15-8836-5_4.

Wang, Songliang, and Renye Wu. 2022. *Exploration of Ecological Civilization: Perspectives and Case Studies (in Chinese)*. Chongqing: Southwest University Press.

Wang, Songliang, Claude D. Caldwell, and Wenfeng Zhu. 2010. Low carbon agriculture: origins, principles and strategies. *Research of Agricultural Modernization* 31 (5): 604–607.

Wang Songliang, Caldwell C. D., Kilyanek S.L., & Sziklay J. (Eds.). 2012. Agroecology. Beijing: China's Science Press.

Wang, Songliang, Claude D. Caldwell, Shannon L. Kilyanek, and Sean M. Smukler. 2019. Using agroecology to stimulate the greening of agriculture in China: a reflection on 15 years of teaching and curriculum development. *International Journal of Agricultural Sustainability* 17:298–311.

Wezel, Alexander, S. Bellon, T. Doré, C. Francis, D. Vallod, and C. David. 2009. Agroecology as a science, a movement and a practice. *Sustainable Agriculture* 2:27–43.

Wezel, Alexander, Philippe Fleury, Christophe David, and Patrick Mundler. 2014. The food system approach in agroecology supported by natural and social sciences: topics, concepts, applications. In *Agroecology, ecosystems, and sustainability*, ed. Noureddine Benkeblia. Boca Raton, FL: CRC Press.

Xu, Zhenci, Sophia N. Chau, Xiuzhi Chen, Jian Zhang, Yingjie Li, Thomas Dietz, Jinyan Wang, Julie A. Winkler, Fan Fan, Baorong Huang, Shuxin Li, Shaohua Wu, Anna Herzberger, Ying Tang, Dequ Hong, Yunkai Li, and Jianguo Liu. 2020. Assessing progress towards sustainable development over space and time. *Nature* 577:72–78.

Zhang, Qian Forrest. 2024. From sustainable agriculture to sustainable agrifood systems: a comparative review of alternative models. *Sustainability* 16:9675.

Chapter 2
The 'What', 'How', and 'Why' of Agriculture and Agroecology

Abstract The challenges of our present agriculture and food systems are consequences of the ideologies of our recent past in agriculture. This chapter critically reviews definitions and technologies of what has been considered 'conventional' agriculture. We consider the consequences and propose a new paradigm for the second Green Revolution under the school of agroecology: Low-carbon agriculture/climate-smart agriculture.

Keywords Conventional agriculture · Low carbon agriculture · Climate smart agriculture · Ecology · Agroecology paradigm · Agroecology education · Sustainable Development Goals

Learning Objectives

After reading this chapter, students should be able to:

1. Compare and contrast definitions of agriculture, including their approach to technologies, and how they have been forged by the teaching/learning disciplines.
2. Expand on the lessons for agriculture in the future from the study of agroecology.
3. Explain the barriers and opportunities involved in implementing low-carbon agriculture/climate-smart agriculture.

****

> Is agriculture a good idea?
>
> —Jared Diamond (1987)

> Conservation is the state of the harmony between the men and the land. By land is meant all things on, over, or in the earth. Harmony with land is like harmony with a friend; you cannot cherish his right land and chop his left land. …. Land is organism.
>
> —Adl Leopold (1966)

The question 'Is agriculture a good idea?' has been addressed by famous scholars like Jared Diamond in 1987 (Diamond 1987) and Clive Ponting in 1991 (Ponting

S. Wang, C. D. Caldwell, *Advanced Agroecology*,
https://doi.org/10.1007/978-3-032-08689-1_2

1991). Their analyses were mostly considered from the view of archaeology and anthropology. Before we address this question from an agroecology standpoint, we first need to have a consensus on what agriculture is from the perspective of ecology. This chapter explores the current patterns and consequences of global agriculture and addresses agroecology's roles in finding a positive way forward.

1 Reflections on Definitions, Technologies, and Disciplines

Our ancestors survived and flourished for approximately 300,000 years using hunting and gathering techniques before gradually adopting new ways of culturing food starting approximately 10,000 years ago. This is approximately the time when agriculture first evolved. Diamond (1987) questioned whether agriculture was indeed better for the livelihood of humankind than what had been successful for 300,000 years. Whatever the answer to that question, it has become apparent recently that the current dominant agricultural production systems are not sustainable due to intensive fossil energy use and an unwavering obsession for pursuing yield and total production goals as the only avenue for feeding the world population. Hence, there is a need to refresh our understanding of agriculture.

For a long time, hunting and gathering formed the stable method to provide food and fibre. Even today, hunting and gathering remains a dominant lifestyle for some peoples such as the Bushmen in South West Africa, Pygmy groups in equatorial forests of Africa, and some aboriginals in Australia. Where hunting and gathering still exists, it provides some advantages over today's agriculture (Ponting 1991). In reassessing the definition of agriculture, it is worthwhile to probe into those advantages as part of reviewing the history of agriculture.

As agriculture developed around 8000 BC, huge changes in human livelihood pattern evolved; it was a major transition in human history. Agriculture stimulated settlement, city establishment, division of labour, politics, and religion and resulted in what some anthropologists call 'the Neolithic Age'. However, it was a gradual evolution rather than an instant change both temporally and spatially. Success for the new agriculture meant that more food was harvested, and more people were fed than with hunting and gathering. Agriculture became the new means of 'successful' living and its benefits accumulated as new and successful methods developed. It was a long-term evolution of agriculture as the means of successful living, and our ancestors slowly gained the knowledge of the basis of agriculture: domestication.

The key to the initiation of successful agricultural methods was the domestication of plants and animals (Caldwell and Wang 2020). Domestication may be defined as:

> The process of establishing a relationship between people and a plant or animal in which the plant or animal is no longer 'wild' but instead lives in association with people, for at least part of its life cycle, and provides some benefit to people.

Domestication began with the gathering of wild plant seed and the subsequent sowing of the seed initially by nomadic groups. Cereal grains such rice and wheat were domesticated first since they are starchy, concentrated packets of food which are easily stored and transported. According to archaeological studies, this initial domestication occurred around 30,000 years ago, well before the first agricultural systems were put in place. However, domestication of plants and animals was a precursor to successful agriculture. It is said that the dog was the first domesticated animal and acted both as the hunters' assistant and as a sentry around dwellings.

Over the centuries of domestication, natural selection of the wild animal traits was replaced by the directed selection for traits which are beneficial to humans. Plants became easier to cultivate and harvest and had larger harvestable parts. Animals were more docile and produced more usable product (meat, milk, etc.).

With this increased reliance on domesticated plants and animals, people became more sedentary and worked more for the cultivation of those crops and animals and relied less on hunting and gathering. Agriculture was then defined as a practice of cultivation of crops and rearing animals, which still is one of the dominant but outdated definitions of agriculture today. Before the practices of cultivation and animal husbandry became established, the process for wild plant and animal domestication to meet the human demand was really a matter of directed adaptation.

In the agricultural evolution processes, the wild traits of plants and animals were domesticated to meet the needs of people, such traits include morphological, physiological, and behavioural adaptations. Potato is a good example; scientists have found wild potato in Chile, Peru, and Bolivia and compared them to cultivated varieties; wild potato is morphologically small (Fig. 2.1) and is also loaded with some toxin which makes the wild potato tubers inedible.

In contrast with plants and animals in natural ecosystem, the morphological, physiological, and behavioural adaptations in agricultural systems are the results of human intervention rather than natural selection.

With domestication and adaptation, agriculture has replaced hunting and gathering as a means of livelihood. In the past 10,000 years, agriculture has evolved from primitive methods through what some people would consider 'traditional' agriculture to what we now call 'modern' or 'conventional' agriculture. In general, people assume that modern methods provide a better livelihood pattern than hunting and gathering, although there are some significant issues regarding human health and environmental integrity. This can be further elucidated by its evolving definitions.

The Oxford Concise Dictionary defines agriculture as the 'science and practice of cultivating the soil and rearing animals' (Oxford Dictionary Editorial Board 2019); this definition is a logical succession from the original understanding of agriculture building on the primitive ideas of tilling soil and feeding animals for human need. However, it does not embrace the function of agriculture's secondary product revolution, which expanded agriculture from solely food and feed to clothing, medicine, and diverse other products resulting from the processing of agricultural feedstuffs. This basic definition does not include the economic complexity of the agricultural sector. Considering agriculture as solely plant and animal production decreases the farmers' motivation for a low-level career. When people have the mindset of only production in agriculture, it negatively affects the politics in

Fig. 2.1 The morphological difference between wild (left, selection of Andean wild potatoes, courtesy: International Potato Center: CIP, Peru, URL: https://www.potatopro.com/news/2021/native-potatoes-forgotten-crop-culinary-boom-and-market-innovation) and cultivated potatoes (right, courtesy: Ms. Zhaojuan Zhang). Note: Almost 200 species of wild potatoes (tuber producing species occur from Southern Chile to the United Sates, but the major wild ancestors of the domesticated potato, Solanum tuberosum, are from Peru and Bolivia around Lake Titicaca)

policymaking process and does not treat agriculture as a formal, basic, and important industry. Therefore, economists have suggested a redefinition of agriculture, expanding agricultural components from food and feed production to fibre, forest, agricultural equipment and materials and their services (Leones et al. 1994).

Meanwhile, in some developing countries, there exists the trend to overstate the economic value of agriculture. China, which is known for its agrarian tradition, seems more focused on the monetary value of agriculture, which can be illustrated by the definition of agriculture in its general agronomy textbook (Zhai 1999):

> Agriculture is the main source of living, and as well as the industrial materials for human beings; And the infrastructure of national economy, even the material sources of foreign exchange earnings via exports for the state.

Such a definition of agriculture perhaps exaggerates the economic value of agriculture and the emphasis on economics must be balanced against the cost of natural resources and the reality of environment deterioration and degradation. It also does not deal with the lack of fairness in many agricultural value chains, particularly, it is far away from the worldwide recognized concept of sustainable agriculture.

2 Lessons of Agriculture and Its Science: The Blind Men Guessing an Elephant

Each discipline relevant to agriculture was originally established in response to problems and opportunities farmers encountered in practice. However, universities have evolved into siloed institutions in which one may say 'Farmers have problems and universities have departments' (Lieblein et al. 2000). There is a disconnection and reductionism in agricultural disciplines that present barriers to solving real-world problems. In practice, those disconnected and fragmented disciplines of agriculture are like 'the Five Blind Men guessing the elephant', each segment considers their analysis of the problem to be reasonably correct according to its particular expertise; however, the true nature of the elephant is not apparent to any of them (Fig. 2.2).

The 'Green Revolution' of the mid-twentieth century is a good example of misunderstanding and lack of communication prevalent in the ideology of conventional agriculture.

The 'Green Revolution' is generally considered to be the period during 1960–1970 when the highly productive cereal crop varieties were bred, and those cereal crops yield per capita and gross production increased dramatically to supply the demand of increasing population. The 'Green Revolution' was characterized by three distinguished cases, in 1962, the first highly productive, dwarf wheat varieties were bred in Mexico by Dr. Norman Borlaug; in 1965, highly productive, dwarf rice variety IR8 was acquired in the International Rice Research Institute (IRRI) in the Philippines; and in 1973, 'Three-line' hybrid rice was first achieved by Professor Yuan Longping in China. As has been well documented, unintended problems emerged from the Green Revolution due to some mistaken approaches to agriculture.

Firstly, *wrong goals and focus.* Despite a significant increase in the productivity of some important crops, negative effects could have been avoided if the emphasis had been placed upon the development of livelihood system strategies rather than a complete emphasis only on yield.

In the name of supporting the increasing population, the three main starchy food crops, rice, wheat, and corn, were placed as the top priority, people ate them three

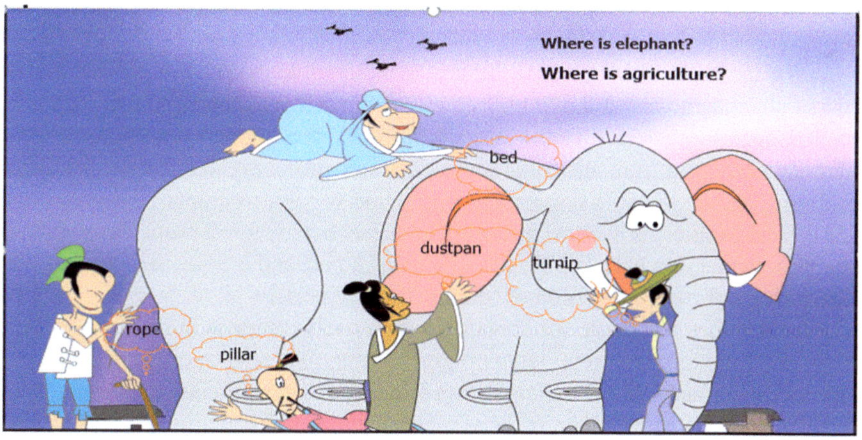

Fig. 2.2 The fragmented disciplines diagnosing agriculture like the five blind men guessing an elephant: the blind men argued heatedly among themselves, but none of them knew what an elephant looked like and agriculture alike

meals a day. Consequently, over time, nutritional balance in people has become a problem. In 2024, micronutrient deficiencies affect 3 billion people. At the same time, farms have become larger, forcing smaller landowners out of business. Those small farmers were not able to afford the cost of land rent and 'improved' agricultural materials.

The understandable desire to increase grain yield blinded the system to other needs and led scientists to equate yield with nutrition, short-term gains with long-term stability, and yield improvement over farm improvement.

Emphasizing plant genetics over almost everything else exploited 'wild' species incorrectly for food security. The broader picture of livelihood systems, soil management, landscape, and biodiversity was lost from the view and community approaches were undermined from the start.

Unfortunately, this fragmented approach remains today in agricultural faculties. Genetic breeding of grain crop varieties is put on the top priority; agronomy tends to have a lower profile and, in many circumstances, breeders both have higher position and salary. Consequently, monoculture of several crops prevails, agrobiodiversity is decreasing both in single crop varieties and field plant diversity dimensions. For the former, since 1900, more than 90% of genetic diversity of crops has declined greatly with the highest extinction rate of 98% of asparagus and the lowest extinction rate of 86% of apples (Thrupp 2000); regarding field diversity, fewer than 100 crops provide 95% of all food, among them, three (rice, wheat, and maize) provide 60% of our food (Paoletti et al. 1992).

In addition, in the Green Revolution, there was an emphasis on *a quick fix*. This was driven both by the severity of the global need for more food and by the funders involved. This resulted in a specialization strategy focusing on single commodities and identifiable problems. In many cases, this approach of producing short-term technical solutions from agricultural sciences undermined long-term sustainability.

It seems that great achievements are always accompanied by great problems in the history of human society development. In many cases, problems appear more often than achievement gains, it is perhaps because humankind is accustomed to using the methods which created the problems to solve the problems. This sociological phenomenon is called 'taking care of the problem at hand' or 'therapy intervention' in medical terms. In the agricultural field, practitioners are apt to mix the methods they learn from agricultural sciences and agronomy to solve the problem, finally producing a whole system problem while solving the superficial problems. The typical example is in conventional plant protection, where a pesticide is applied to a pest, neglecting both the system impact of a pest and a pesticide, creating the old ideology of plant protection of 'eliminating the pest'.

Another place where we see damage to the overall system is in soil health. With the goal of maximizing crop yield per area of land, the use of more and more synthetic fertilizer and other chemical inputs often results in degradation of soil health. Fertilizer, pesticide, and other artificial inputs damage soil structure and biotic components causing soil contamination. Soil contamination leads inevitably to food safety problems, which is an essential part of food security, especially in those fast-developing countries in which agriculture is the primary economic driver.

Another manifestation of the Green Revolution which has caused some unintended problems is capitalization of agriculture and vertical integration. Several huge companies control large portions of the agricultural industry worldwide; for example, the world's grain food controllers are ADM, Cargill, Bunge, and Louis Dreyfus who handle 60% of the grain market. In the European community, 10 retailers control 90% of foodstuffs. This situation affects food distribution and profit allocation to farmers and hence equity and fairness in the value chains.

A continuing issue in agriculture is gender bias. It is a common bias for people within and outside of the agricultural sector, to assume that "men" and "farmers" are synonymous. This is also wrong and has caused great harm from this inequality issue. People are gradually recognizing the central role of women in food production, but the-male-dominated research system is still slow to adjust. Women are the main labour force and educator in rural and agricultural fields, and women's knowledge is often the keeper of indigenous knowledge. It is key to secure sustainable food production and distribution and pass it on to subsequent generations. Unfortunately, the role of women is underestimated and almost undermined by globalization and the spread of science and technology (Momsen 2007).

To restructure the current unsustainable agricultural system, it is not enough just to adopt novel technologies, such as information technology, biotechnology, nanotechnology. Success will not come from following an old path; on the contrary, it will very likely result in the following causal chain: increased high-technology leads to more commercialization of agriculture production, leading to more farmer poverty, more energy input, more land and water degradation, and more food security and safety problems. In the end, staying with the old path leads to continued lack of sustainability.

As we have analysed, the first generation of Green Revolution was not truly 'green'; rather, it had misplaced priorities that have led world agriculture in an unsustainable direction. Therefore, if we need a new Green Revolution in agriculture that can quickly respond to the requirement for doubling of available food by 2050 and environmental crises, it should be truly 'green' (Dahlberg 1979; Holt-Giménez and Altieri 2013). Ideologically, the true 'green' Green Revolution points to ecological and systematic thinking aimed at both food security and food sovereignty. Disciplinarily and scientifically, agroecology will be right choice as a guideline and model to the goal of a new Green Revolution (Holt-Giménez and Altieri 2013; Horlings and Marsden 2011).

3 An Agroecology Paradigm: Towards the Totally New Green Revolution

To rescue the current unsustainable agricultural system, it is not enough just to adopt the new or novel technologies, such as information technology, biotechnology, nanotechnology, and so on, if so, we will never get out of it if we follow the old

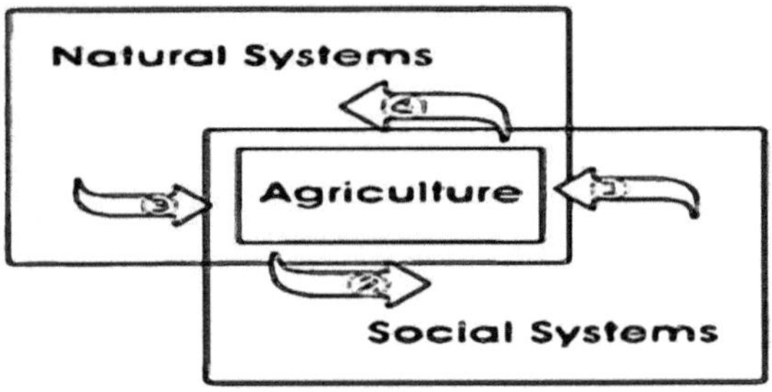

Fig. 2.3 Agriculture exists at the interface between natural systems and social system

obsolete route; on the contrary, it will be very likely to result in the following causal chain: the more high-technology applied, the more commercialization of agriculture production, the more farmer poverty, the more energy input, the more natural degradation, and the more food security and safety problems, finally, the less sustainability. In this regard, sustainable development of agriculture will be doomed to true slogan.

Since we now understand that our traditional definition of agriculture is not only incomplete but also has become part of the problem, it is imperative that there be a redefinition of agriculture.

3.1 Convergence in Redefining Agriculture

In general, people have a lack of appreciation and understanding of agriculture; we have failed to recognize that the basic function of agriculture is the interface between nature and society (Dahlberg 1979, Fig. 2.3). Agriculture was developed to meet the demands and objectives of societal systems; over time, it has been overlooked as a component of natural systems, resulting in the degradation of natural systems.

Human history is determined by our relations with natural ecosystems, a fact that is dramatically substantiated by our present climate change crisis. Agriculture is not a standalone discipline but rather a convergence of multiple disciplines.

For the past two decades, the authors have been using the following convergent definition of agriculture for teaching, research, and extension purposes and it has expanded our opportunities for learning and the application of holistic approaches to food-based problems (Caldwell 1996; Wang 2005):

> Agriculture is the science, art, politics, and sociology of changing sunlight into healthy and happy people.

This definition directly addresses the essence of agriculture, i.e. human health and happy life. Surely this is why people invented agriculture. Solar energy utilization is the first important thing in agriculture and food systems, and plants, rather than animals and humans, are the only organisms that can utilize solar energy and thus form the basis for all life (Fig. 2.4).

Students who approach the study of agriculture, using only a reductionist scientific approach, will be frustrated by the endeavour. As a convergent discipline, agriculture is a systematic enterprise that creates, builds, and organizes knowledge in

Fig. 2.4 The conceptual model of the new definition of agriculture by the authors (Sketched by Ms. Yiyang Wang, a UBC undergraduate student in 2022)

the form of testable explanations and predictions. The scientific lens for considering agriculture is already complex since its functional niche is in ecology, a holistic and integrative science.

As *art*, agriculture is something that is created with imagination and skill, has innate beauty, and makes important ideas concrete as it links natural ecosystems with human-built landscapes. Agriculture acts as the interface between nature and society.

As a *sociological* endeavour, agriculture involves the systematic study of the development, structure, interaction, and collective behaviour of organized groups of human beings. The sociological lens requires us to consider agriculture as a public good sector for human health and ecosystem integrity.

As a *political* study, agriculture deals with the structure, organization, and administration by the state. It is common for governments (both local and federal, capitalist, and socialist) to utilize policies to manage agriculture and its rural associations, in some cases providing a 'safety net' for farmers, such as price floors, stabilizing reserves and supply-demand management. Markets are affected by various incentives or controls to provide targeted products or perhaps ecosystem services.

These aspects of science, art, sociology and politics in agriculture are not only intertwined but also interdependent. What is produced when and where is certainly affected by soil, water,** and climate but also is usually governed by political priorities and concrete policies which impact both the social and ecological fabric of communities. Farmers must understand the science behind production and the complex interactions of plants, animals, microbes, and the environment. Successful farmers integrate this knowledge with the specifics of the social, economic, and political dimensions to produce the livelihood components of ecosystem services, the integrated nature-based welfare system that sustains our existence.

Therefore, as the above definition indicates, agriculture goes far beyond production and the current understanding of agriculture. Agriculture involves the full value chain, affecting economy, environment, and health; it is ultimately controlled by consumer behaviour. Participants in sustainable value chains (those that form the web from producer to consumer) must address communication, equity, and fairness to find the 'sweet spot' of profitability, ecosystem integrity, and human health.

Therefore, redefining agriculture, using the convergence of disciplines, is the prerequisite for an essential agriculture and food system.

Sociologically, this involves reconnecting producer and consumer, linking the food system to human health, and educating ourselves particularly as consumers in the full value chain to develop a lifestyle of low-carbon consumption.

Ecologically, acceptance of the redefinition involves reconnection of plant and animal production and land use planning, i.e. closing the loop on the nutrient movements and cycles.

Politically, there is a need for stabilization of primary producers and their associations which will lead to rural empowerment with spinoff into social stability.

To utilize the new definition of agriculture, there is a need to break down the boundaries in teaching and learning in agricultural sciences. This means embracing the natural, social, and economic knowledge of agricultural and food systems as a whole. The realization of such integration is nested in agroecology. There is a need for education in agroecology, linking Sustainable Development Goals (SDGs) and stimulating reinvention of a sustainable food system locally, regionally, and globally. As the core vehicle for the new agriculture for rebuilding our agri-food systems, agroecology is positioned in the niche between agricultural sciences, rural sociology, and ecology (Fig. 2.5).

3.2 *Why Are We Learning Ecology?*

In agroecology, ecology is the core component. The term ecology was coined in the nineteenth century by a German animal scientist Ernst Haeckel; he merged the two Greek words, 'Oikos', meaning 'home 'or 'house', and 'Logos', meaning 'study 'or 'science' to 'Oikologos', it was then translated to ecology in English and basically

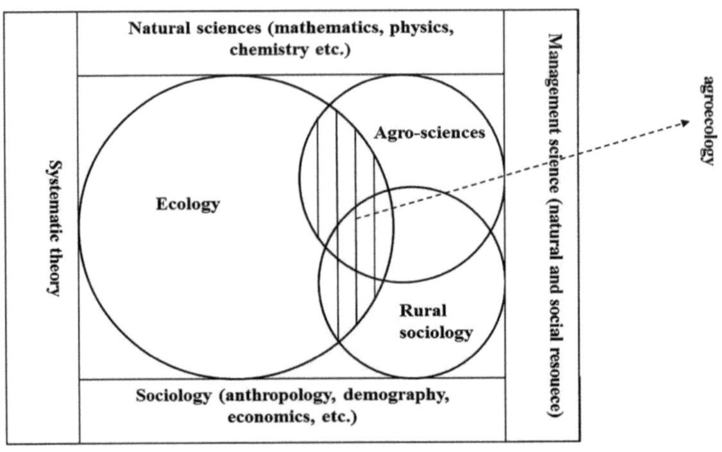

Fig. 2.5 The niche of agroecology (Wang et al. 2012)

was defined by Ernst Haeckel as the science of organisms interacting with their living environment (Odum 1953). Haeckel considered both the social and environmental crises that evolved gradually after the initiation of the industrial revolution in Europe. He was concerned with the reductionist scientific approach lacking any link between people and the environment. Hence, the mission of ecology from its inception is twofold, to define and to solve the ecological crises encountered in the contemporary world.

The disciplinary conceptual framework of ecology and agroecology is to locate ecology as inter- and cross-disciplinary of social and natural sciences. E. P. Odum used 'the link between natural and social sciences' as the subtitle of his famous ecology textbook in its second edition (Fig. 2.6, Odum 1975).

Apart from acting as the bridge between social and natural sciences, there is a need for an integrated systematic theory for ecology to realize the link, i.e. transitioning reductionism to systematic thinking.

Since economics plays a major role in our daily life, finding a platform for dialogue between ecology and economics is crucial to develop sustainable transitions and to facilitate eco-economics as a sustainable development science. In this regard, it is important to review the gaps between two disciplines.

In 1942, the young ecologist, Raymond L. Lindeman proposed the concept and principles of food chains in an aquatic ecosystem (Lindeman 1942). The food chain and food web concepts illustrate that food is the link between the hierarchical organisms and to their environment in an ecosystem; as trophic level increases, the

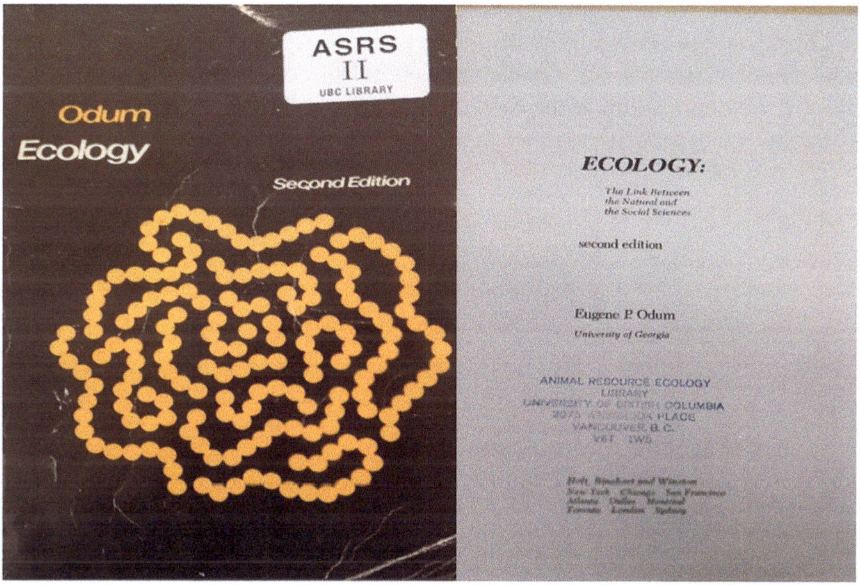

Fig. 2.6 The cover page of E. P. Odum's second edition *Ecology* with the subtitle of 'the link between natural and social sciences' (Scanned from UBC library by Songliang Wang at Spring of 2019)

number of individuals, biomass of populations, and energy reduce, forming a pyramid-like structure. This concept of trophic levels illustrates how energy reduces when transforming from soil through plants (producers), animals (consumers), and microorganisms (decomposers), finally to higher level consumers. When this concept is expanded to agri-food systems, energy reduces in the same way when transforming from the arable land(soil), through crops and livestock, finally to humans. However, this contrasts with financial flow in economics; monetary value is expected to be magnified as much as possible when transforming from the arable land (soil), through crops and livestock, finally to human, forming an inverted pyramid, as comparatively displayed on the right side of Fig. 2.7. Based on this analysis, the general answer to the question 'why are you learning ecology?" is to put the economic pyramid within the boundary of the ecological pyramid. The logic is obvious, the human economy is a subset of the economy of nature (Czech 2019). Economic expectation and development in society represent human desires but relies on and is limited by natural resource and environment; it cannot transcend the resource limitation and environmental capacity; otherwise, ecological crises are inevitable.

Agriculture and food systems are an appropriate platform for a dialogue between ecology and economics. Consider the trophic level theory in agri-food. Economically, energy, matter, information, and value transform from land via crop, livestock, and processed food to finally become a commodity/product that produces monetary income. Therefore, the economy is based on the food chain of agriculture, starting from land (soil), through growing crops and feeding livestock, and finally forming a traded commodity. Therefore, whatever the economist wants value-added, the industry begins with agriculture, specifically with crops (plants) through the process of photosynthesis, called 'primary production'. This in turn results in secondary production, processing, and development of various commodities and products. In other words, agriculture is the fundamental service for the economy of society. Using the same logic, the farmer is the real engine of human economy just like plants are the real producers of the economy of nature (Czech 2013, 2019;

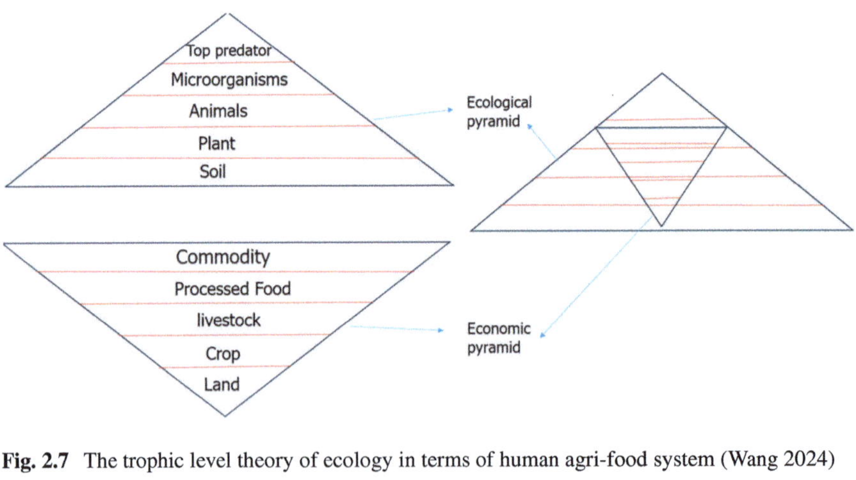

Fig. 2.7 The trophic level theory of ecology in terms of human agri-food system (Wang 2024)

Gardes-Landolfini et al. 2024). In other words, 'Money grows on a tree', not 'a tree grows by money' (Czech 2010). Integrating ecology into agriculture to form a discipline called agroecology is a logical vehicle for ecology and economics to dialogue and to realize true sustainable development (Wang 2011).

3.3 The Basic Tenets of Agroecology

Agroecology is an acronym of agricultural ecology, literally, applying the principles and methodologies of ecology in agricultural studies.

Agroecology appeared as an academic term in Italy in the1920s, interpreted as applying ecological thinking for protecting the soil under various cropping systems (Klages 1928). During the soil Dust Bowl of the 1930s in Western USA, government, scientists, and farmers realized that soil is key to production agriculture and agroecology became the disciplinary term defined as ecology in agriculture. Agroecology became a focus for soil conservation by using ecology approaches and traditional intercropping (Odum 1953). With Odum's modern ecology emerging in the 1950s, ecologists became active in agricultural research, bringing to bear ecological principles and methodology (Odum 1953). It was Azzi, an Italian professor in 1956 who first began to lecture his undergraduate students in an agroecology course; in his lectures, he defined agroecology as the marriage of ecology and agriculture (Francis et al. 2003). However, agroecology did not become a formal discipline or science until the world environmental conference in 1972; after that time, agroecology was gradually established in the faculties of agriculture and food in land-grant universities of USA.

The first book named *Agroecology* was published by Cox and Atkins (1979); it advocated for ecological thinking on the world food security system. Subsequently, M.A. Altieri, at the University of California in Berkeley, started to build the agroecology discipline in 1983 by defining agroecology as an ecology of alternative agriculture, a new model of modern agriculture that was proposed to substitute for conventional agriculture (Altieri 1983). Besides Altieri, another famous ecologist, Stephen R. Gliessman from the University of California in Santa Cruz proposed that sustainable agriculture is the ideal goal of world agriculture development and hence defined agroecology as ecology of sustainable agriculture (Gliessman 1990). This was in response to WCED's formal SD concept release in 1987 (WCED 1987).

The systematic thinking and interdisciplinary approach of agroecology developed from ecology. Such a framework of agroecology not only unites the compartmentalized disciplines in the agri-food system but also defines a totally new Green Revolution for the next generation of agriculture, the ideal end point of which is sustainable agriculture.

The following tenets are nested in the theory and practice of agroecology:

A positive, participatory, and inclusive approach to developing food systems of the future. To their detriment, people in modern society (consumers) are separated from production agriculture (producers) and its products. All stakeholders in agricultural value chains have the right and opportunities to rebuild the system. The new systems require a participatory system to be re-established with an intrinsic and positive attitude.

- **Complying with Natural Ecosystem Processes:** From the disciplinary framework of ecology, the natural ecosystem processes broadly constitute energy flow, material cycling, and information transformation. Since soil is the fundamental infrastructure of agri-food systems, soil conservation should be the very top priority.
  - ***Low-Carbon Use*****:** Understanding energy flow in agroecosystems is key to reducing carbon emissions. Low-carbon agriculture or climate-smart agriculture is an important goal of sustainable agriculture. Therefore, low-carbon use becomes indispensable for agri-food system rebuilding (Cederlof 2016). This refers to both reduction of emissions and soil carbon sequestration (Yang et al. 2003; Lal 2004).
  - ***Water Use Efficiency:*** Water is one of the key limiting factors of food production in agroecosystems (Wang 2012); it has large impacts on global and regional food security and safety (Brown and Halweil 1998; Jia et al. 2010) and water use efficiency should be promoted in ecologically compatible agri-food system (Sarangi et al. 2004; Li et al. 2011).
  - ***Agrobiodiversity:*** Soil health underground and agrobiodiversity above the soil are two of the most important aspects for securing sustainability in agri-food systems. Both have been threatened by our prevailing industrialized agriculture model.
  - ***Animal Welfare***. Animal welfare is gaining attention in the agroecologically transformed agri-food systems. Animal welfare is not only related to ethical treatment to animals as a production component and type of our food but also has a close connection to food quality and human health.
- **People-Centred Agri-Food System Re-establishment**: Food is defined by more than just production; it has both ecological and social aspects. People involved in the agri-food system should be healthy physically and happy psychologically. As such, the distribution of resources and profits within the system should be established with equity and fairness, with sensitivity to roles of women and youth, fairness in value chains within economic subsystem, and most important respect for transdisciplinary knowledge gained through experience and scientific method investigations.

  Based on the abovementioned tenets, we have not only bridged ecology to agricultural sciences for agroecology but also set up a connection with rural sociology to define the niche of agroecology as illustrated in Fig. 2.5. The goal is to build agroecology as a realistic, information-based integration of science, sociology, and politics for designing sustainable agri-food systems; in other words, to find a sweet spot linking health, environment, and economy (Fig. 2.8).

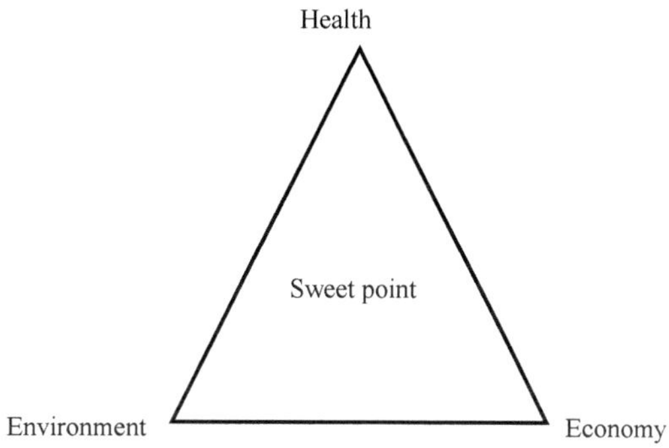

Fig. 2.8 Using agroecology to find the "sweet spot" of balance among environment, economy, and human health of agri-food system (Caldwell and Wang 2023)

3.4 Functional Approaches to Agroecology

Within the realm of agroecology, there have evolved new agriculture models such as organic agriculture, regenerative agriculture, low-carbon agriculture or climate-smart agriculture, and ecological agriculture, all together amended targeting sustainable agriculture (Zhang 2024).

- **Organic Agriculture (OA):** OA was proposed in the United States in the 1920s, the focus of which was on producing food without synthetic chemicals and, more recently, without genetically modified seeds. It focuses primarily on the process and certification of methods for food production based on specific standards for environmental and organic practices. Generally, the long-term OA practice can promote sustainable development of agriculture by decreasing input of synthetic chemical compounds, using intercropping and green manure to preserve soil and water quality and reducing nonrenewable resource use, while decreasing greenhouse gas production (Zinna et al. 2005; Gomiero et al. 2011). One limitation of OA is that it is primarily driven by consumer concern for food safety rather than other social factors, such as equity and fairness. While there is a base group of smaller OA producers, OA has also become large scale on an industrial level with corporate profit becoming the driving force. In the larger farms, managing ecosystems holistically and sustainably is not needed to maintain certification.
- **Regenerative Agriculture (RA):** The concept of RA originated in the 1950s in the United States (Dahlberg 1996) but gained significant worldwide attention recently. Based on the idea of restoring, conserving, and enhancing natural resources by minimizing waste and creating self-sustaining cycles, RA prioritizes soil health and enrichment. The farming approach of RA is very holistic, focusing on soil regeneration, biodiversity, restoration, and ecosystem health improvement. Common techniques may include cover cropping, no-till farming, complex rotations, and managed grazing. It is important to note that the goal of regenerative agriculture is not just food production but also the improvement of ecosystem services. One of the limitations of RA is that it is still a problem solution-oriented model, lacking systematic ideology and integrating approaches may lead to other natural and social problems in the long run (Tittonell et al. 2022).

- **Climate-Smart Agriculture (CSA):** CSA was first proposed in 2009 as an alternative agricultural model to address food security under global climate change, which aimed at agricultural mitigation of carbon emissions and adaptation to global warming to achieve the goal of food security for the increasing population (Lipper et al. 2014). The pivotal practice of CSA embraces less or zero-till farming, traditional intercropping with cover crops and forages, agroforestry, and organic amendment. However, like regenerative agriculture and organic agriculture models, CSA has been criticized for its problem solution-oriented framework: therefore, it focuses on technical fixes, politically top-down and corporate-led approaches (Zhang 2024). It also lacks a specific set of criteria and a farmer-centred marketing mechanism.
- **Ecological Agriculture (EA)**: Both literally and disciplinarily, EA refers to using thinking and principles of ecology to design and manage agroecosystems. EA was proposed in 1970 by a United States soil scientist William A. Albrecht (Mercies 1978) and defined in 1981 by British agronomist Martha Kiley-Worthington as (Kiley-Worthington 1981):

  > The establishment and maintenance of an ecologically self-sustaining low input, economically viable, small farming system managed to maximize production without causing large or long-term changes to the environment or being ethically or aesthetically unacceptable. It respects the natural cycles of each product to avoid negative impacts to the local ecosystem.

  EA complies with principles of ecology and targets transforming the large-scale industrialized agriculture model to small-scale ecologically compatible agriculture, which was adopted and practised by smaller countries of the eastern Europe countries, such as Romania and Poland (Ion et al. 2014; Krochmal-Marczak et al. 2013). Almost at the same time, ecology and agroecology were introduced in China as disciplines; by doing so, the ecologists in China attempted to use principles of ecology to slow the speed of the nation's industrialized, chemical-based agriculture. They named the new agricultural model as ecological agriculture (Wang and Shi 2023). Such EA in China has been practised more than 40 years and has achieved significant gains. EA, at least in China, appears to be a sound pathway to transform industrialized, chemical-based and capital-intensive agriculture to sustainable agriculture.

  In 2010, Olivier de Schutter, UN food rights committee officer submitted a report named 'Agroecology and Right to Food' to the UN commission and it was subsequently approved (de Schutter 2010). The report suggests that agroecology is the best pathway to lead world agriculture to sustainable agriculture by means of systematic thinking and ecological approaches to solve the problems exhibited by current conventional agriculture. Referring to the expanding definition of agroecology by Wezel et al. (2009), i.e. as a science, as a practice, and as social movement (Wezel et al. 2009), the report suggested using these multiple meanings of agroecology to unify its function in rebuilding the future agri-food system. In this context, when in its application and practice, field agroecology is equivalent to EA. In a nutshell, the report claimed EA would be a workable model for future sustainable agriculture (Chap. 10 explores this further).

- **Sustainable Agriculture (SA):** The proposition of SA originated from the concept of sustainable development (SD) by the Brundtland Commission of the United Nations in 1987 (WCED 1987). After several experiences in SA programs such as LISA (low input sustainable agriculture) and SARD (sustainable agriculture and rural development), in 1991, the FAO provided a comprehensive SA definition (Xu et al. 1997):

  > The application of husbandry experience and scientific knowledge of natural processes to create agriculture and agri-food systems that are economically viable and meet society's need for safe and nutritious food and vibrant rural communities, while conserving or enhancing natural resources and the environment.

  In the minds of many, SA has become an ideal for future agriculture; however, it is both ideal and vague, necessitating a concrete and enforceable practice to make it a reality (Dahlberg 1991). Agroecology or ecological agriculture seems to fit best (Francis et al. 2011; Wezel et al. 2014).

3.5 Agroecology and Climate-Smart Agriculture

When comprehensively considering the critical challenges of the world's agri food system, climate change, food insecurity, natural resource and environmental degradation, and social inequality, one needs a combination of ecological and climate-smart agriculture (previously called low-carbon agriculture, aiming at sustainable agriculture under the school of agroecology (Wang et al. 2010). The essential component of modern fossil fuel-based agriculture is high-carbon consumption and emission agriculture.

CSA is a basic and important step towards sustainable agriculture since it is key to mitigation and adaptation to climate change. Under the school of agroecology, to establish a system of low-carbon input and output, farmers utilize an integrated farm management strategy (IFaM). The intent is to integrate yield and marketing with soil health, biodiversity, water conservation, and animal genetics/health with crop production in terms of matter and energy flows. The intent is to find a sweet spot (Fig. 2.8).

Under such an integrated farm management strategy, after the cropping agenda has been decided at the farm level, there are several sub-strategies that are implemented (Fig. 2.9), i.e. integrated strategies for management of fields, crops, water, soil, weeds, pests, and plant/animal nutrition (Wang and Caldwell 2013).

On an operational level, there are adaptations that need to be made to realize climate smart (low carbon) goals of agriculture (Wang and Shi 2023):

- Multidimensional Land Use

  Multidimensional land use technology is an ecological/agricultural technology system that employs the principles of ecological niche and biological succession to develop and fully utilize natural space resources. In the vertical space, there is

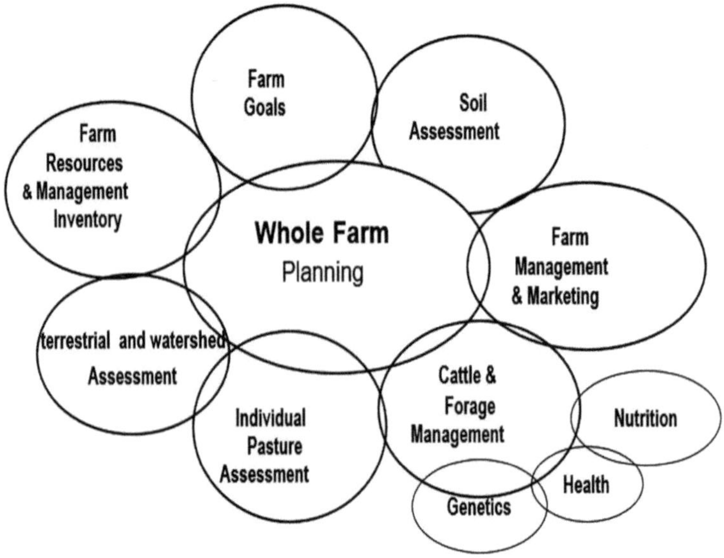

Fig. 2.9 The integrated management strategy for a climate smart-based farm program

a recognition of the different characteristics and growing requirements of various species to allow for appropriate combination of planting and aquaculture systems. This establishment of multiple species in a common habitat ensures efficient utilization of resources; this approach is also known as 'stereoscopic agriculture'. For example, the traditional 'agroforestry system' uses forest resources and varying environments in the canopy to plant crops. These may include herbal plants, edible fungi, and even livestock and poultry to form a composite system of agriculture, forestry, and animal husbandry. This increases biological diversity to fully and efficiently use natural resources, reduce the input of materials and energy from outside the system, and finally form an efficient agricultural production system.

- Multistage Utilization
  Multistage utilization of matter and energy technology refers to the use of the principles of ecological food chain and ecological pyramid to form a new food chain, decreasing losses and increasing efficiency in the system. For example, currently in Southern China, fish farming in rice fields is one of the simplest ecology-based agriculture models, using the water space between rice plants to add "fish" food links to achieve multiple goals: Firstly, in order to ensure the survival of fish, rice fields do not apply chemical fertilizers and pesticides; secondly, fish will eat harmful organisms on rice plants, and fish waste can fertilize the field, which can reduce the input of fertilizers and pesticides. Thirdly, the concentration of fish in the rice field promotes the nutrient reduction reaction of the paddy soil, greatly reducing the emission of methane and therefore alleviating the greenhouse effect.

- Integrated Pest Control
  The integrated pest control technology system uses the principles of ecological niche and inter-species relationships to comprehensively apply cultivation technology, physical means, biological control, and rational use of pesticides to recover crop losses caused by pests and achieve substantial pesticide reduction, not only to maintain the health of agricultural production environment, but also to ensure the safe supply of urban and rural food and important agricultural products and cut the carbon emissions substantially.

3.6 Agroecology Education Linking to Sustainable Development Goals

At present, there is a disconnect between the implementation of the United Nations Sustainable Development Goals (SDGs) and agriculture as it is practised. Agriculture suffers from being fragmented, poorly understood, and having a poor image both with society and young people. Over the last 10,000 years, agriculture has been a key positive force for health, development of society, culture, and happiness across the globe. Unfortunately, within the past 100 years, agriculture has been reduced to science and capitalism and has become a significant source for many of humanity's social and environmental problems. At present, agricultural practices contribute considerably to greenhouse gas problems, economic disparity, and the rural-urban split.

Who should be blamed for the abovementioned wrongness of agriculture, 'wrong definition', 'wrong technology', and 'wrong ideology'? Obviously, it is education's responsibility to deliver the holistic thinking and knowledge to stakeholders of agriculture and food. Unfortunately, current prevailing compartmentalized and fragmented disciplines in the universities and colleges have failed to fulfil the function. Consider an agriculture science college or faculty as an example to demonstrate the paradox. The discipline set in the college or faculty is usually organized as follows:

- **Plant genetics and breeding,** a popular and superior discipline, focusing on genetics-based research and high yield variety development.
- **Crop cultivation**, working on the regulation of environmental factors for high yield realization of high yield bred varieties.
- **Plant protection,** its goal is to eliminate the pests in the system.
- **Soil chemistry and plant nutrition,** focusing on the mineral ion absorption mechanism and fertilizer development and application.
- **Plant biotechnology**, advocating genetically modified crops and other gene editing plant development and application as main route for a new green revolution.

Agriculture faculties at universities across the globe reflect this academic fragmentation and are failing to serve communities in terms of solving complex problems linked to food. Our food systems have complex, often interdisciplinary challenges. Universities are traditionally fragmented in their structure, divided into faculties, departments, and subdepartments, often with no clear interdisciplinary cooperation. This has hindered the ability of universities to address complex issues practically. The use of assembly line education in agriculture does not work; it appears that, while students may learn many of the parts of agriculture, they often miss key information links and lack the personal 'navigation software' to address the problems of the twenty-first century. One key aspect of our food system is value chains that have been greatly affected by the fragmentation of agriculture. Supply chains should facilitate the ultimate provision of food which has positive effects on health, environment, and the economy. However, our supply chains are highly competitive when what is needed is cooperative value chains.

Since the nineteenth century, universities have adopted the factory assembly line model for education. Alvin Toffler in his 1970 book *Future Shock*, notes that 'mass education was the ingenious machine constructed by industrialism to produce the kind of adults it needed' (Cuban 2014, 2022) . This was a time before large highways, vaccines, knowledge of DNA, TV, mass media, global warming, space travel, and a population of 9 billion people. Our education system, especially for pre-university, tends to instil information but throttle innovation and problem-solving skills. University faculties of agriculture continue the fragmentation of education so that the final product (student graduate) may have all the bits and pieces of education but not truly be educated. In other words, there is a disconnect between the graduate and the problems he/she must face in terms of the food systems of the twenty-first century.

The convergence of the definition of agriculture leads to the necessary application of agroecology. In terms of education, agroecology is the logical mechanism for equipping people with the knowledge, principles, and interdisciplinary approaches that are presently separated in our current agricultural curricula. We have been using this definition of agriculture as core ideology in teaching, research, and extension aspects of agroecology for two decades in China and Canada. The result has been the development of new agriculturalists with an expanded knowledge and appreciation of agriculture and the global challenges. Our goal continues to be the promotion of knowledge, understanding, and action that evolves from our new definition and the implementation of agroecology globally.

By redefining agriculture, we dig into the root cause of the disconnect between humans and our environment. The opportunity is to teach the reinvention of agriculture's natural essence and to connect with the 'sweet spot' of sustainable agriculture between economy (economically viable'), society ('meeting society's need for safe and nutritious food and vibrant rural communities'), and ecology ('conserving or enhancing natural resources and environment'). The convergent approach of agroecology facilitates the integration of knowledge of nature, society, and economy and emerges as a multidisciplinary approach representing the resurfacing of socially, environmentally, and culturally sustainable productive systems. Through the promotion of ecological processes beneficial for the soil, water conservation,and biodiversity management, agroecology provides guidelines for agricultural transformation: efficient use of water, reduced use of agrochemicals, improved soil health, mitigation and slowing climate change, doubling farmer income, and providing food for everyone in an equitable manner.

It is of primary importance that agricultural education in agroecology be linked to SDGs. The pedagogical approach should be focused on problem-solving, particularly focused on the reinvention of sustainable food systems locally, regionally, and globally. Instruction at the university level needs methods and approaches of interdisciplinary convergence. Curriculum (broadly defined as content and methods) must accommodate and integrate the knowledge and characteristics of multiple disciplines, sciences, art, sociology, and politics, in order to open up our minds and embrace an original, innovative spirit of sustainable development. This also must spillover into research and development. No time for us to wait, we need collective action right now.

References

Altieri, Miguel A. 1983. *Agroecology*. Berkeley: University of California Press.

Brown, Lester R., and Brian Halweil. 1998. China's water shortage could shake world food security. *World Watch* 11 (4): 10–18.

Caldwell, Claude D. 1996. *Agricultural ecosystems* (Unpublished course manual). Truro: Nova Scotia Agricultural College (now Dalhousie University).

Caldwell, Claude D., and Songliang Wang. 2020. *Introduction to agroecology*. Singapore: Springer.

Caldwell, Claude D.,and Songliang Wang. 2023. Agroecology lecture for Sino-Canadian 2+2 program in the campus of Fujian Agricultural and forestry University

Cederlof, Gustav. 2016. Low-carbon food supply: the ecological geography of Cuban urban agriculture and agroecological theory. *Agriculture and Human Values* 33:771–784.

Cox, G. W., and M. D. Atkins. 1979. *Agricultural ecology: an analysis of world food production systems*, 721. San Francisco, CA: W. H. Freeman.

Cuban, Larry. 2014. *Schools as factories: metaphors that stick*. https://larrycuban.wordpress.com/2014/05/08/schools-as-factories-metaphors-that-stick. Accessed January 21, 2025.

Cuban, Larry. 2022. *Larry Cuban on school reform and classroom practice: schools as factories: metaphors that stick*. https://larrycuban.wordpress.com/2014/05/08/schools-as-factories-metaphors-that-stick. Accessed January 09, 2025.

Czech, Brian. 2010. *The trophic theory of money*. Steadystate.org/the-trophic-theory-of-money/. Accessed January 21, 2025.

Czech, Brian. 2013. *Supply shock: economic growth at the crossroads and the steady state solution*. Gabriola Island, BC: New Society Publishers.

Czech, Brian. 2019. The trophic theory of money: principles, corollaries, and policy implications. *Journal & Proceedings of the Royal Society of New South Wales* 152 (part 1): 66–81.

Dahlberg, Kenneth A. 1979. *Beyond the green revolution: the ecology and politics of global agricultural development*. New York: Plenum Press.

Dahlberg, Kenneth A. 1991. Sustainable agriculture: fad or harbinger? *Bioscience* 41 (5): 337–340.

Dahlberg, Kenneth A. 1996. A transition from agriculture to regenerative food systems. *Futures* 26 (2): 170–179.

De Schutter, Olivier. 2010. *Agroecology and right to food*. Report presented at the 16th Session of the United Nations Human Rights Council[R/OL]. Human Rights Council of the United Nation, (2010-12-20). http://www.srfood.org/en/reportagroecology-and-the-right-to-food. Accessed November 2024.

Diamond, Jared. 1987. The worst mistake in the history of the human race. *Discover* 8 (5): 64–66.

Francis, C. A., Porter, P. 2011. Ecology in Sustainable Agriculture Practices and Systems. *Critical Reviews in Plant Sciences* 30: 64–73

Francis, C., G. Lieblein, S. Gliessman, T. Breland, N. Creamer, R. Harwood, L. Salomonsson, J. Helenius, D. Rickerl, R. Salvador, M. Wiedenhoeft, S. Simmons, P. Allen, M. Altieri, C. Flora, and R. Poincelot. 2003. Agroecology: the ecology of food systems. *Journal of Sustainable Agriculture* 22:99–118.

Gardes-Landolfini, Charlotte, William Oman, Jamie Fraser, Mariza Montes de Oca Leon, and Bella Yao. 2024. *Embedded in nature: nature-related economic and financial risks and policy considerations*. IMF Staff Climate Note 2024/002. Washington, DC: International Monetary Fund.

Gliessman, S. R. 1990. *Agroecology: researching the ecological basis for sustainable agriculture*. New York: Springer.

Gomiero, T., D. Pimentel, and M. G. Paoletti. 2011. Environmental impact of different agricultural management practices: conventional vs. organic agriculture. *Critical Reviews in Plant Sciences* 30:95–124.

Holt-Giménez, Eric, and Miguel A. Altieri. 2013. Agroecology, food sovereignty, and the new green revolution. *Agroecology and Sustainable Food Systems* 37 (1): 90–102.

Horlings, L. G., and T. K. Marsden. 2011. Towards the real green revolution? Exploring the conceptual dimensions of a new ecological modernisation of agriculture that could 'feed the world'. *Global Environmental Change* 21:441–452.

Ion, R., I. R. Emilia, N. C. Georgeta, et al. 2014. Study on the recent trends in ecological agriculture from Romania. In *SGEM International Multidisciplinary Scientific Geo-Conference EXPO-Proceedings*, 393–398. https://doi.org/10.5593/SGEM2014/B53/S21.052.

Jia, Shaofeng, Shijun Lin, and Aifeng Lv. 2010. Will China's water shortage shake the world's food security? *Water International* 35 (1): 6–17.

Kiley-Worthington, M. 1981. Ecological agriculture. What it is and how it works. *Agriculture and Environment* 6 (4): 349–381.

Klages, K. H. W. 1928. Crop ecology and ecological crop geography in the agronomic curriculum. *Journal of America Social Agronomy* 20:336–353.

Krochmal-Marczak, B., M. Dykiel, and B. Bienia. 2013. Management of quality control in the ecological agriculture in Poland. In *International Conference: Hradec Economic Days 2013—Economic Development and Management of Regions*. Poland, iii.

Lal, R. 2004. Soil carbon sequestration to mitigate climate change. *Geoderma* 123:1–22.

Leones, Julie, Gerald Schluter, and George Goldman. 1994. Redefining agriculture in interindustry analysis. *American Journal of Agricultural Economics* 76 (5): 1123–1129.

Leopold, Aldo. 1966. *A sand country almanac with essays on conservation from round river*. New York: Ballantine Books.

Lieblein, G., C. Francis, and J. King. 2000. Conceptual framework for structuring future agricultural colleges and universities in industrial countries. *The Journal of Agricultural Education and Extension* 6 (4): 213–222.

Lindeman, Raymond L. 1942. The trophic-dynamic aspect of ecology. *Ecology* 23:399–418.

Li, Lingling, Gaobao Huang, Ren-zhi Zhang, Bill Bellotti, Guangdi Li, and Kwongyin Chan. 2011. Benefits of conservation agriculture on soil and water conservation and its progress in China. *Agricultural Sciences in China* 10 (6): 850–859.

Lipper, L., P. Thornton, B. M. Campbell, T. Baedeker, A. Braimoh, M. Bwalya, P. Caron, A. Cattaneo, D. Garrity, and K. Henry. 2014. Climate-smart agriculture for food security. *Nature Climate Change* 4:1068–1072.

Mercies, J. R. 1978. Towards ecological agriculture. *Futuribles* 16:479–484.

Momsen, Janet Henshall. 2007. Gender and agrobiodiversity: introduction to the special issue. *Singapore Journal of Tropical Geography* 28:1–6.

Odum, Eugene P. 1953. *Fundamentals of ecology*, 546. Philadelphia, PA: W. B. Saunders Company.

Odum, Eugene P. 1975. *Ecology: the link between the natural and the social sciences*. 2nd ed. New York: Holt, Rinehart and Winston.

Oxford Dictionary Editorial Board. 2019. *Agriculture*. Oxford: Oxford Dictionary Editorial Board.

Paoletti, M. G., D. Pimentel, B. R. Stinneff, and D. Stinneff. 1992. Agroecosystem biodiversity: matching production and conservation biology. *Agriculture, Ecosystems and Environment* 40:3–23.

Ponting, Clive. 1991. *A new green history of world-the environment and the collapse of great civilizations*. New York: St. Martin's Press.

Sarangi, A., C. A. Madramootoo, and C. Cox. 2004. A decision support system for soil and water conservation measures on agricultural watersheds. *Land Degradation and Development* 15:49–63.

Thrupp, L. A. 2000. Linking agricultural biodiversity and food security: the valuable role of agrobiodiversity for sustainable agriculture. *International Affairs* 76 (2): 265–281.

Tittonell, P., V. El Mujtar, G. Felix, Y. Kebede, L. Laborda, R. Luján Soto, and J. de Vente. 2022. Regenerative agriculture—agroecology without politics? *Frontiers in Sustainable Food Systems* 6:844261. https://doi.org/10.3389/fsufs.2022.844261.

Wang, Songliang. 2005. Information technology: toward the sustainable management of agroecosystem (in Chinese). *Agriculture Network Information* 8 (4–7): 12.

Wang, Songliang. 2011. Historical review of agronomy both as a discipline and a specialty and its reframing: taking agroecology as its core theoretical subject (in Chinese). *Chinese Journal of Eco-Agriculture* 19 (6): 455–460.

Wang, Songliang. 2012. Using ecology thinking reconstructing traditional agronomy: role of production ecology (in Chinese). *Chinese Journal of Applied Ecology* 23 (8): 2031–2035.

Wang, Songliang. 2024. *Why are we learning ecology?* Presentation in the seventh world ecosummit congress, Decembner 19–23, 2024, in Zhengzhou, Henan, China.

Wang, Songliang, and Claude D. Caldwell. 2013. Reconstructing traditional agronomy with ecological thinking: taking agroecosystem management as the core applied subject (in Chinese). *Chinese Journal of Eco-Agriculture* 21 (1): 39–46.

Wang, Songliang, and Shengxu Shi. 2023. Developing Chinese ecological agriculture acts as a vital pathway of the state-specifically agricultural modernization in China: a historical review of ecological agriculture movement in China. *Chinese Journal of Eco-Agriculture* 31 (8): 1184–1193.

Wang, Songliang, C. D. Caldwell, and W. Zhu. 2010. Low carbon agriculture: origins, principles and strategies (in Chinese). *Research of Agricultural Modernization* 31:604–607.

Wang, Songliang, C. D. Caldwell, S. L. Kilyanek, and J. Sziklay, eds. 2012. *Agroecology*. Beijing: China's Science Press.

Wezel, Alexander, S. Bellon, T. Doré, C. Francis, D. Vallod, and C. David. 2009. Agroecology as a science, a movement and a practice. A review. *Agronomy for Sustainable Development* 29 (4): 503–515.

Wezel, Alexander, Marion Casagrande, Florian Celette, Jean-François Vian, Aurélie Ferrer, and Joséphine Peigné. 2014. Agroecological practices for sustainable agriculture. A review. *Agronomy for Sustainable Development* 34:1–20.

World Commission on Environment and Development (WCED). 1987. *Our common future*. Oxford: Oxford University Press.

Xu, Cheng, Zeng Xiaoguang, and Wang Erda. 1997. *Introduction to sustainable agriculture [in Chinese]*. Beijing: China's Agriculture Press.

Yang, Dawen, Shinjiro Kanae, Taikan Oki, Toshio Koike, and Katumi Musiake. 2003. Global potential soil erosion with reference to land use and climate changes. *Hydrological Processes* 17:2913–2928.

Zhai, H. Q. 1999. Introduction to agriculture. In *Toward 21st century textbook plan series*. Beijing: Higher Education Press.

Zhang, Qian Forrest. 2024. From sustainable agriculture to sustainable agrifood systems: a comparative review of alternative models. *Sustainability* 16:9675.

Zinna, Yuri L., Rattan Lala, and Dimas V. S. Resck. 2005. Changes in soil organic carbon stocks under agriculture in Brazil. *Soil and Tillage Research* 84:28–40.

Chapter 3
Agroecosystems and Their Services

Abstract Agroecosystems, managed ecosystems, accommodate both economic and ecological productivity of agriculture. Nested within the principles of agroecology is a relatively new concept in ecology called agroecosystem services. This chapter delineates the primary understanding of ecosystem services relating to human welfare and societal sustainability and proposes a framework to monetize agroecosystem services in the current economic value chain and policy system: a Payment for AGroEcosystem Services (PAGES).

Keywords Ecosystem · Ecosystem services · Agroecosystem · Agroecosystem services · Payment for Agroecosystem services (PAGES)

Learning Objectives

After reading this chapter, students should be able to:

1. Explain the concepts of ecosystem, ecosystem services, and agroecosystem services.
2. Classify agroecosystem services and describe their evaluation methods.
3. Define the links between agroecosystem services and global food security.
4. Discuss options for a viable policy for implementing PAGES.

****

> It's the demand side, stupid.
> —James Carville

> Eventually we'll realize that if we destroy the ecosystem, we destroy ourselves.
> —John Salk

Agroecosystems, managed ecosystems, accommodate both economic and ecological activities of agriculture, which together impact agroecosystem services. Ecosystem service is a relatively new concept in ecology; however, it has become a controversial issue both in academic circles and society in general in the past two decades. Ecosystem service assessment and evaluation have been initiated in

S. Wang, C. D. Caldwell, *Advanced Agroecology*,
https://doi.org/10.1007/978-3-032-08689-1_3

response to environmental degradation due to agricultural policy and market failures. These are coupled with the concepts of 'Sustainable Development', 'sustainability', and 'Sustainable Development Goals (SDGs)'.

Agroecosystems provide vital services to support human society. The provision services are undervalued by current marketing mechanisms, and supporting, regulating, and cultural services are routinely overlooked. There is little credence given to the concept of monetary compensation to farmers even though this would be a strong and reasonable motivation to farmers to produce and enhance those ecosystem services. On the contrary, farmers around the world have suffered the effects of the long-term insufficient economic recognition of providing those services. This has directly led to rural abandonment and chemical-born food safety problems because of the pressure to produce more production from less income. This in turn jeopardizes agroecosystem landscape integrity and its service delivery. It is time to deliver monetary compensation for agroecosystem services provided by farmers. In this chapter, after addressing the general issues of ecosystem services and agroecosystem services, the authors explore a realistic framework of Payment for Agroecosystem Services (PAGES) coupled with a feasible policy suggestion. This will be supported by documentation of recent publications and citation of the experiences and lessons of PAGES implementation worldwide.

1 Ecosystems and Their Services

'Ecosystem' is a principal concept of ecology initiated by the British ecologist Arthur G. Tansley in his famous academic paper 'Use and Abuse of Vegetational Concepts and Terms' in 1935. After addressing the significance of the linkage between plants and their habitats, he proposed the new term to encompass such significant interactions; he said in a general statement (Tansley 1935):

> The whole system (in the sense of physics) including not only the organism-complex, but also the whole complex of physical factors forming what we call the environment of the biome-the habitat factors in the widest sense. Though the organisms may claim our primary interest, when we are trying to think fundamentally, we cannot separate them from their special environment, with which they form one physical system. It is the systems so formed which, from the point of view of the ecologist, are the basic units of nature on the face of the Earth.

Therefore, such whole systems, ecosystems, can be defined as the organisms living in a certain area interacting mutually and with the living and the non-living features of their environment within the area. The proposition of 'ecosystem' was a revolutionary event in the evolutionary history of biology, because since then, people are advised to look at the relationship between organism and their environment rather than the separate individual organisms. Such interaction is not static; instead, it is dynamic because both organisms and environment are dynamic (Fig. 3.1).

The dynamic ecosystem is characterized by matter, energy, and information flows among the organisms and between organism and environment. Ecologists call

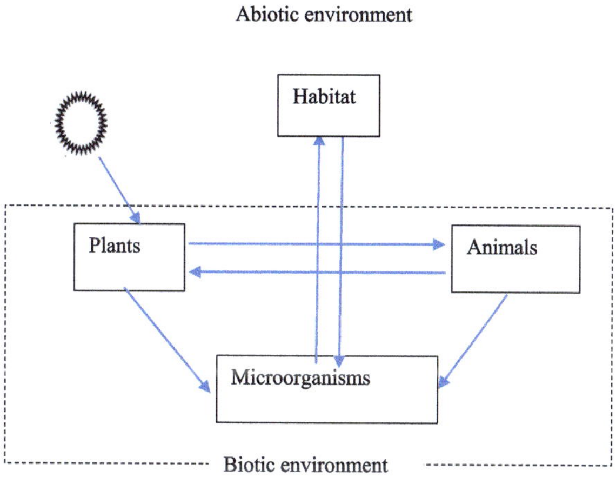

Fig. 3.1 A dynamic ecosystem model shown by plants, animals, and microorganisms interacting with each other and with biotic and abiotic environment at any time to gain a dynamic equilibrium

these flows ecological processes and functions of ecosystems, which result in the 'services' benefiting the ecosystem itself and humans within it. This is the genesis of the term 'ecosystem services'.

The term ecosystem service (ES) was first proposed by Ehrlich and Ehrlich in 1981 and was popularized by Gretchen C. Daily (1997) and Costanza et al. (1997); the contribution of Gretchen C. Daily was to connect ecosystem services (ES) to natural capital (Daily 1997) and the contribution of Robert Constanza was to quantify the economic values of ES to the global ecosystem (Costanza et al. 1997). Then an organization called Millennium Ecosystem Assessment (MEA) formally categorized ES into four categories (Fig. 3.2), i.e. provisioning services (PS) such as food, fibre, fodder, forage, forest, and fuel (the so-called the six F's, in most literature distinguished as ecosystem goods, we refer as PS hereafter), regulating services, e.g. carbon dioxide sequestration and water purification, supporting services such as soil formation and nutrient cycling, and cultural services such as scenic landscape for the purpose of aesthetic and spiritual fulfilment (MEA 2003, 2005a, b).

In short, ecosystem services are the benefits people obtain from ecosystems, which are key to human existence, survival, and development.

However, at the beginning of the ecosystem service proposition, the pioneers distinguished provisioning services from the other three categories, naming the former as ecosystem goods and the latter as environmental services (Simoncini 2009). In general, the modern free-market system, which was constructed under neo-liberal economics, has recognized and valued the ecosystem goods (the six F's mentioned above). Unfortunately, the other environmental services have not been valued similarly.

With the ecological crises deepening, there has evolved a common understanding that ES are closely coupled with human welfare (MEA 2005a, b). More and more scholars assert that ES provide almost the entire welfare of mankind, meaning that almost everything we use originates from a natural ecosystem, without which human civilization cannot exist and is therefore not sustainable (Daily et al. 1997).

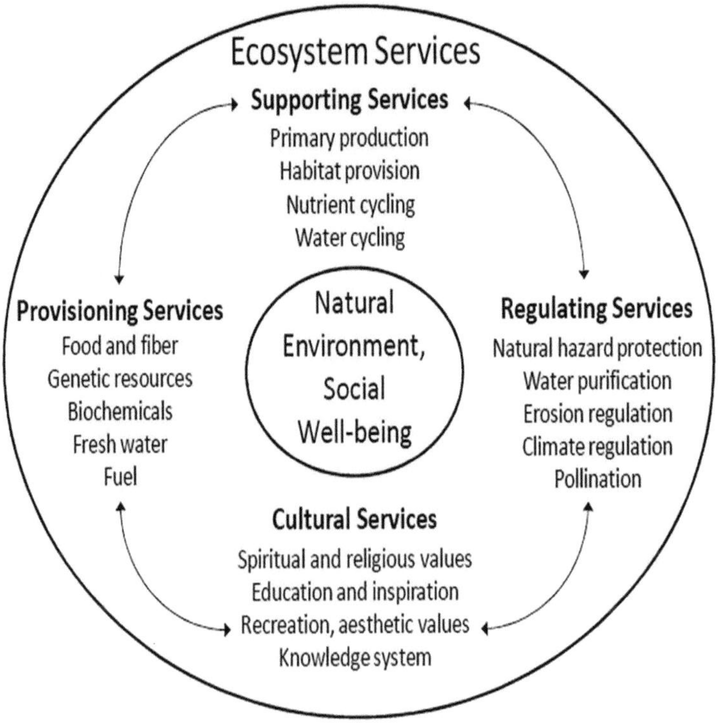

Fig. 3.2 The four components of ecosystem services

As depicted in Fig. 3.3, human welfare has two sources, from nature and from society, the former refers to ecosystem services, including ecosystem goods and environment services, presently known as 'natural capital' under eco-economics; the latter consists of money, mind, and spirit-based support with the new name of 'social capital' or human-made capital. However, such human-made capital has a unique source, nature (Daily et al. 1997).

Wang (2007) classified the ecosystem goods into seven categories, foodstuff, site views, power, recreation, materials, fuel, and others, such as indicators, predators, and pollinators. These have been valued and monetized in some way in various industries by the modern marketing paradigm. However, the method and degree of monetizing remain controversial (Fig. 3.4).

Among these, medicine in the materials category and all the other services are usually heavily underestimated. One example is Chinese Sweet Wormwood (Fig. 3.5), from which the 2015 Nobel laureate Madam Tu Youyou and her group devoted two decades for the exaction of artemisinin to cure the world's malaria; even now its value is still much undervalued. There remains a great deal of research to do for scientists and practitioners of ecosystem service implementation; however, the even larger question that remains is: can we put a monetary value on these services besides the ecosystem goods or provisioning services (Matulis 2014)?

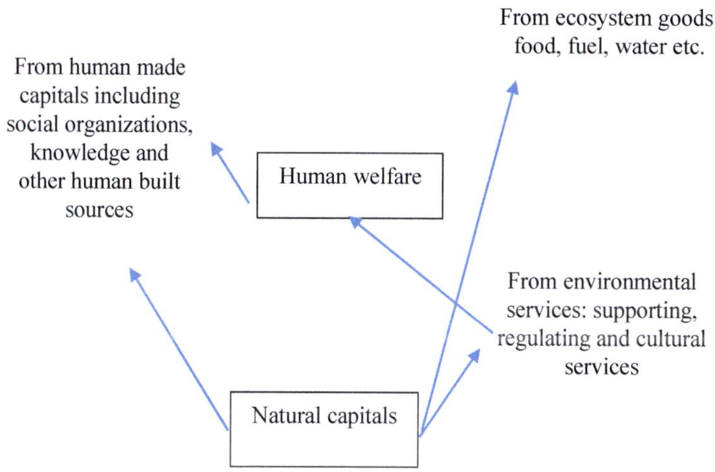

Fig. 3.3 Does all human welfare emanate from ecosystem services?

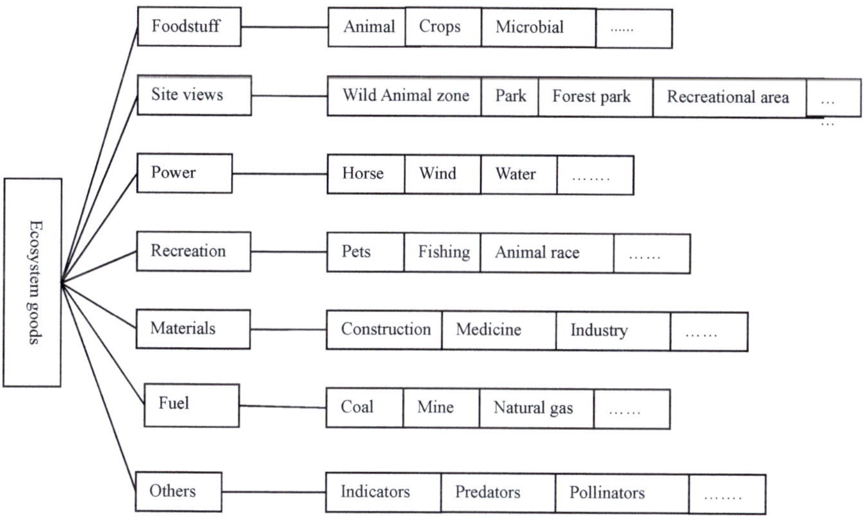

Fig. 3.4 The ecosystem goods classification (Wang 2007)

In addition to the Millennium Ecosystem Assessment (MEA, www.millennium-assessment.org) which initiated the ecosystem service assessment process globally, several international and intra-continental organizations have explored the question with substantial applications (Zhang et al. 2019). Among these, the Economics of Ecosystems and Biodiversity (TEEB, www.teebweb.org) focuses on 'making nature's values visible' and 'mainstreaming the values of biodiversity and ecosystem services into decision-making at all levels' since 2010; the Intergovernmental Platform on Biodiversity and Ecosystem Services (IPBES, www.

Fig. 3.5 Chinese Sweet Wormwood, an herbaceous weed has medical effect on malaria

ipbes.net) was established as 'an independent intergovernmental body' 'to convene a plenary meeting to determine modalities and institutional arrangements for' 'Organizing Focal Points of ecosystem service assessment' also since 2010; and the Ecosystem Services Partnership (ESP, www.es-partnership.org) was set up as a cross European organization to integrate the ES assessment into the decision-making since 2015. Since 2015, all these initiatives have been able to help promote and advance ecosystem services research towards sustainable development, especially the United Nations Sustainable Development Goals (UN SDGs, sustainabledevelopment.un.org), SDG#15 with the theme of life on land towards the protection, restoration, and promotion of sustainable use of terrestrial ecosystems, sustainable management of forests, combating desertification, and halting and reversing land degradation and prevention of biodiversity loss (Wood and DeClerck 2015; Zhang et al. 2019).

To put ecosystem services into the mainstream policymaking process, a dialogue between ecology and economics is imperative. Ecology teaches people how to understand the ecosystem structure and processes, which are the prerequisite for comprehension of ecosystem function and the resultant ecosystem services. It is then possible for ecosystem service to be valued in a practical and accurate manner, using a rational formula that can be integrated into policy and capitalized (Fig. 3.6).

The ES valuing process is the most important part of the above mechanism. Knowing the ecosystem structure and processes is fundamental to value ecosystem services. Several eco-economic valuing methods may be used (Table 3.1), including market price-based approaches, cost-based approaches (consisting of the replacement cost method, the restoration cost methods, shadow project method,

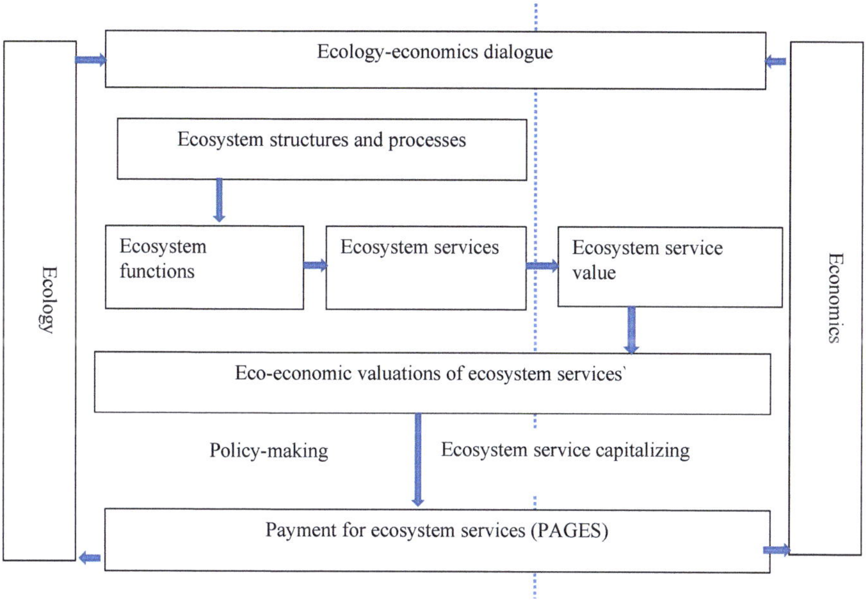

Fig. 3.6 The mechanism for ecosystem services to be recognized and capitalized

Table 3.1 Ecosystem service classification and their valuing methods

Value classifications	Valuing methods
Direct market value	(a) Market price-based approaches (b) Cost-based approaches • The replacement cost method • The restoration cost method • Shadow project method • Opportunity cost method • Carbon tax method
Indirect market value	(a) The travel cost method (b) The hedonic pricing method (c) The willingness to pay and accept method

opportunity cost method, and carbon tax method) for the direct market value of ES, the travel cost methods, the hedonic pricing method, and the willingness to pay and to accept method for indirect market value of ES (Fig. 3.7).

There are some application cases of The Payment for Ecosystem Services (PES). The earliest recognized case was in the USA, the Catskill/Delaware Watershed program in New York in 1997, with a payment budget of 1–1.5 billion dollars for farmers instead of a water cleaning engineering budget of 6.0–8.0 billion dollars, resulting in an effective and cost-saving program in which farmers were rewarded and motivated to protect water quality for the urban public (Rosa et al. 2004; Naeem et al. 2015).

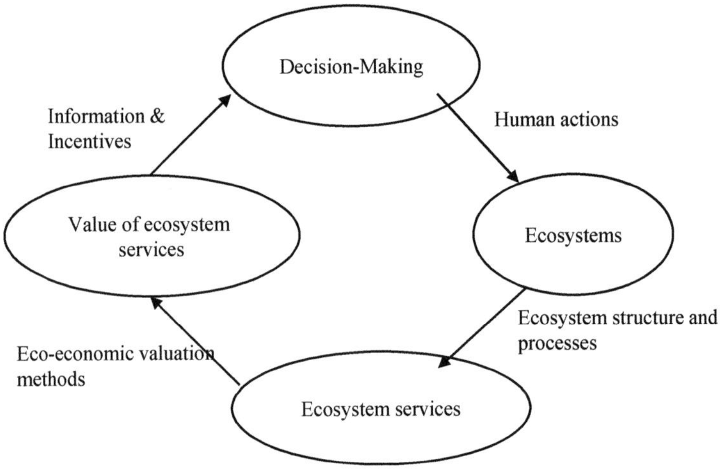

Fig. 3.7 The ecosystem service valuing processes

Another famous and large-scale case was in China, which, since 1999, has implemented the Natural Forest Conservation Program (NFCP) and The 'Grain for Green' (GTGP, also called 'Sloping Land Conversion Program, SLCP'). This has been a huge success. Other examples include one in Costa Rica (Castro et al. 1997) and Mexico's National Forest Commission (CONAFOR) (Zhang et al. 2019).

Even though the abovementioned PES cases relate to agriculture, they are not purely agriculture because they do not connect directly with food production (Ramírez-Suárez et al. 2024).

2 Agroecosystems and Their Services

Agroecosystems are best understood as the unit of study of agroecology, which looks at agricultural production systems in terms of ecosystem prosperities: e.g. stability, resilience, disturbance regime, stocks and flows of nutrients and energy, information transmission, and niche dynamics.

Basically, there are three ways to understand agroecosystems: structure, ecological process, and function. Firstly, agroecosystems are a part of larger natural ecosystems; similar to natural ecosystems, the components of an agroecosystem are twofold: organisms and their environment. However, in the case of agroecosystems, humans act as agency objects rather than simple organisms; the environmental category is also narrower in space and crosses over with society. This makes the structure of agroecosystems much simpler in terms of species and landscape diversity but more complex in terms of social and economic interactions. Secondly, similar to ecosystems, solar energy is the activating power of agroecosystems, without which, agroecosystems are not able to function, maintain, and develop (Conway 1986). The

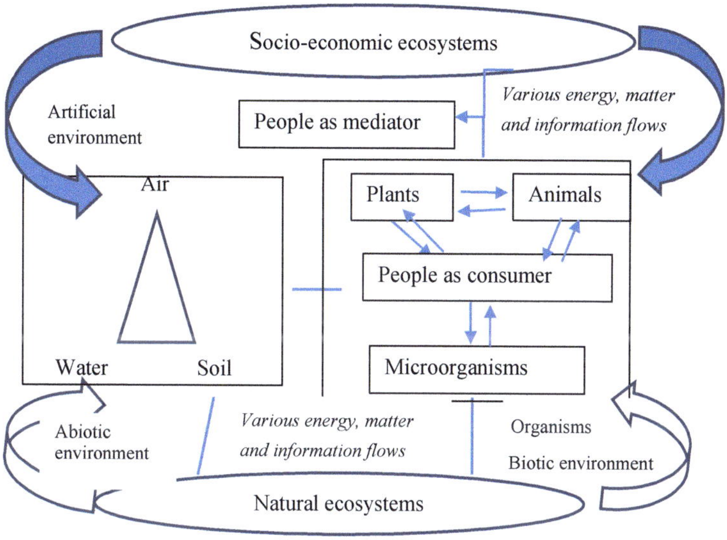

Fig. 3.8 The components and structure of an agroecosystem, indicating that solar energy is still a starting power of agroecosystems as natural ecosystem is, and natural system offers services and socioeconomic one may provide disservices to agroecosystem

nutrient cycle in agroecosystems is usually limited by artificial external inputs according to demand for food and fibre. Agroecosystems are more utilitarian and less resilient compared to ecosystems. Thirdly, unlike agroecosystems, the goal of natural ecosystems is stability, which functions to keep the ecosystem sustainable and incidentally provides ecosystem services to human society. The goal of agro-ecosystems, however, is distinctly to provide ecosystem services to support people's living and development.

The above comparisons result in the agroecosystem model in Fig. 3.8.

Agroecosystem services in general can be interpreted by the multi-functionality of agriculture (Daelemans et al. 2023). The direct provisioning services of agriculture (the 6'f''s, food, fibre, fodder, forage, forest, and fuel) are marketed and rewarded by prevailing economics. Meanwhile, agriculture provides much more in terms of supporting, cultural, and regulating services. These impacts include such difficult to monetize services such as agrarian soil and cropping space which supports key organisms which are key to global biodiversity. Furthermore, agricultural soils sequester carbon; conserve water and stabilize local micro-climates. The cultural services of agriculture are ubiquitous and celebrated: traditional farming systems, governance, and technologies of historic agriculture have had a huge effect on modern agriculture and society; those are best exhibited by the world agricultural cultural heritages listed and propagated by United Nations. The six 'F', Field, Farmer, Food, Family, Fair, and Future, are shown in Fig. 3.9 to characterize the abovementioned indirect values of agriculture.

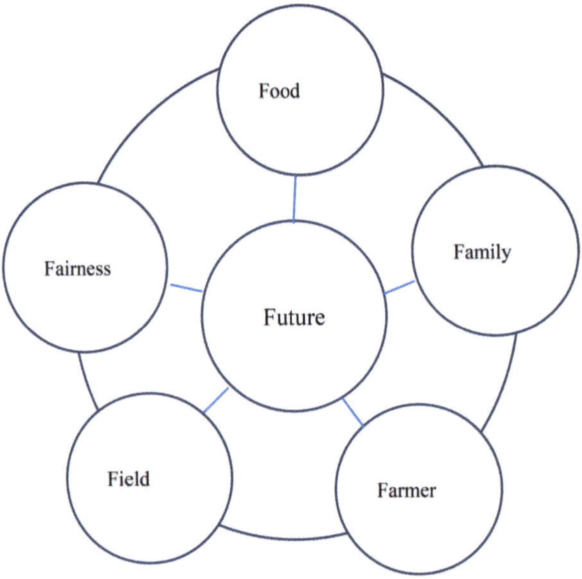

Fig. 3.9 The six 'F' of agriculture characterizes the agroecosystem services

- **Field**: Fields are nature's gift to life on earth; the principal feature of fields is soil that formed over millennia. Soil is endowed with physical, chemical, biological, and morphological properties and characteristics that form an unconsolidated mineral and organic matter on the immediate surface of the earth. Soil serves as a natural medium for plant growth and the plants (designated as crops in agricultural systems) provide fundamental food for humans. If we acknowledge the definition of agriculture as proposed by Caldwell and Wang (changing sunlight into human's healthy and happy life), the plant is the core, then soil is the core, then field is the core.
- **Farmer**: It is the farmer who gets up to work at sunrise and retires at sunset to facilitate that change of sunlight into people's health and happiness; hence, it is incumbent on the rest of us to learn and respect and reward them. We need to acknowledge that farmers are taken for granted world over. This is despite farmers being the basic building block of agriculture and food systems both in mindset and physical niche. It is time to re-learn and re-evaluate the significance of the farmer worldwide by taking the concept of ecosystem service as a vehicle to embody farmer's true value (FAO 2007; Kull et al. 2015).
- **Food**: Food is the fundamental material for human survival and development; there is a Chinese saying, 'Food is the paramount necessity of the people', but most of time people take food for granted. Many of the 8 billion food consumers in the world need to change attitudes towards food.
- **Family:** Farmers who produce food form families, food consumers form families too, and families make up the people and world: in this regard, the family is

the bridge between persons, between communities, between people, and the environment. The earth is a family in terms of food.

- **Fairness:** One of the important properties of an agroecosystem is equality, meaning that every person involved in the system should share the outcome and resources of the system appropriately (Conway 1987; Babin 2015). However, the current agriculture and food system discriminates against the farmer, as is evidenced by the fact that the farmer shares less and less along the economic chain of agriculture since 1910. Taking coffee as an example, coffee farmers share only 5–10% of the coffee value chain, and export and import licence owners have 10% in their pocket, shipping and processing businessmen own 55% of the profit of a cup of coffee, and finally 25% is distributed to the supermarket (Basic 2024).
- **Future.** The future is undoubtedly the most important, but if we are not able to handle the other 5 'F' properly, our future is risky. Future, in an agroecology sense, is equivalent to sustainable development in a sociological sense; it is time to make the sustainable food system visible and implementable.

In addition to the abovementioned two levels of multiple functionalities of agriculture, agriculture needs to be further understood in the following four aspects:

- **Production**: The primary objective of agriculture is to produce food and fibre. That is why the water cleaning project of New York City since 1997 and the Chinese Grain to Green project in terms of payment of ecosystem services since 1999 (see the first section of this chapter) were not truly agroecosystem service payment events. In this regard, the agroecosystem is a common part of the natural and social system with human participation, which is a comprehensive system with agricultural production as the means and food production as the driving force.
- **Livelihood**: The second objective of agriculture is to provide meaningful livelihood. Beyond providing the primary physical needs of food and fibre, part of the role of agriculture is provision of a safe, green environment in which people are able to live and prosper. Agriculture, in its best form, provides an aesthetically favourable landscape to nurture not just health but also happiness. Therefore, it is desirable for farmers to plan and manage their production aesthetically to benefit both physical and emotional health.
- **Ecology**: Meanwhile, agriculture should be ecologically compatible meaning that crops and livestock are linked through circulation of nutrients and energy within the farm as much as possible. The farmstead should also share space with wild plants and animals, in other words, maintaining and building biodiversity. Hong Qian used 'LIVE AND LET LIVE' to describe such harmonious and symbiotic ecosystems, agroecosystems, and social ecosystems (Hong 2012).
- Life: Ultimately, agroecosystems are life support systems and should have a holistic approach to life supporting both quantity and quality of life.

The abovementioned four categories of agriculture are very dependent on ecosystem services; for instance, a meal is a feat of ecosystem services (Fig. 3.10), indicating that the provisions of these meals resulted from many ecosystem services, including natural pest control by beneficial insects, insect pollinators (about

Fig. 3.10 A Chinese meal to show life through agriculture. (Courtesy: Claude D. Caldwell in 1998)

Fig. 3.11 The *Russula* and Wild bamboo shoots are two delicious dishes on the Chinese dining table from the undomesticated natural ecosystem. (Courtesy: Songliang Wang)

90% of commercially grown field crops, citrus and other fruit crops, vegetables, and nut crops currently depend on honey bee pollination services that in some cases could readily be provided by native pollinators living in natural ecosystems), maintenance of soil fertility, purification of water, and moderation of climate. Most of those meals are harvested from agroecosystems; however, some meals on the dining table in modern times are still collected from natural ecosystems, such as the Russula and wild bamboo shoots, as shown in Fig. 3.11. Food, by its very nature, should be appreciated and cherished.

Table 3.2 The value of the agroecosystem services of Fujian Province of China (Zhu 2011)

Categories of agroecosystem services in Fujian Province	Economic value ($\times 10^8$ Yuan)	In percentages (%)	Values per ha (yuan/hm^2)
Marketing product value	1182.740	5.203	9537.189
Ecotourism	1125.000	4.949	9071.595
Art, educational, and scientific value	331.214	1.457	2670.787
Climate regulating	1263.357	5.558	10,187.251
Air freshening	687.335	3.024	5542.419
Water cleaning	2219.303	9.763	17,895.659
Soil conservation	15,862.056	69.779	127,905.905
Resisting natural disaster	61.006	0.268	491.929
Total	22,732.001	100.000	183,302.732

As mentioned previously, agroecosystem services are not limited to those meals on the dining table; they also constitute environmental services. Agrarian soil sequesters carbon; in addition, of course, crops also absorb carbon dioxide as they grow. All agroecosystem services need to be evaluated and capitalized to an appropriate degree to encourage and stimulate the stakeholders, particularly the farmers, to practice and provide ecosystem services. However, for the most part, the value is not fully recognized, let alone monetized. In 2011, an assessment of the agroecosystems services in Fujian province of China showed unexpected results (Table 3.2).

As the data in Table 3.2 shows, the total agroecosystem services value is 22,732 × 10^8 Yuan, accounting for 183,302 Yuan per ha land, and the provisioning services value is merely 9537 Yuan per ha, accounting for 5% of the total value. This implies that the farmer share is undervalued. A 10% share is not a fair and sufficient amount for the farmer. The major value that the farmer has created resides in environmental services, i.e. climate regulating, air freshening, water cleaning, soil conservation, and resisting natural disasters, which account for nearly 90% of the value of agroecosystem services. Among them, soil conservation occupies nearly 70% of the total agroecosystem services, reinforcing the fact that soil is the most important resource for agriculture and food production. Soil is bound to be degraded if provisioning services of agriculture are underestimated and less capitalized by the current market. In this regard, assessing agroecosystem services properly and compensating it appropriately to the farmers will be a crucial means to improve agricultural systems, particularly to conserve soil for the next generations.

The above case suggests strongly that monetization of agroecosystem services could be a very good leverage to improve the management of agroecosystems. Using the ecosystem services concept, the mechanisms of agroecosystem management are shown in the flowchart in Fig. 3.12. There are four mechanisms to operate agroecosystem services concepts illuminating agroecosystem management.

- **Ecological Mechanism:** Agroecosystem management strategy should first be built on the understanding of ecological processes of agroecosystem service formation (Conway 1985; Gliessman 2004); therefore, investigation into functional

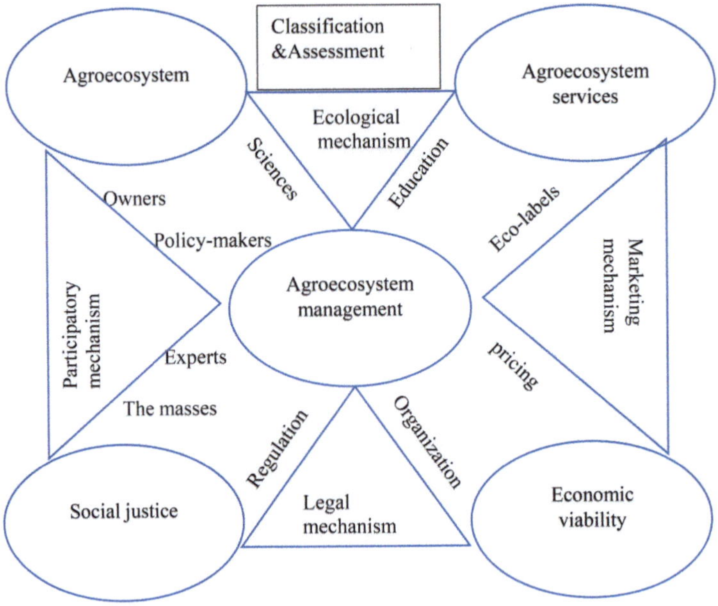

Fig. 3.12 The mechanisms of ago-ecosystem management. (Adapted from Zhu et al. 2010)

processes of agroecosystems and education to deliver the knowledge and foster the necessary talents will be key processes.

- **Participatory Mechanism:** One of the most important properties of the agro-ecology approach to ecosystems is equity, the fair distribution of resources and benefits within agroecosystems (Conway 1987). To realize this goal, the property owners, together with scientists (experts), policymakers, and communities need to be involved in policymaking.
- **Marketing Mechanism:** After the assessment of agroecosystem services, the value should be capitalized step by step; the marketing mechanism then is utilized; eco-labelling and pricing are two keywords in this process before the payment policy is made.
- **Legal Mechanism:** The legal mechanism is the last barrier to achieving the objectives of appropriate distribution of the benefits of agroecosystem services and sustainable management of agroecosystems. The role of law is to establish decision-making procedures related to the management of agroecosystems, not to impose absolute standards.

3 Designing a PAGES for Farmers: A Framework

As already shown, farmers share an inappropriate portion of the current value chain of the agricultural economy, which has become a stubborn problem for rescuing and transferring agriculture and food systems from unsustainable to sustainable status.

The proposition of ecosystem service concept might offer a framework and vehicle to reverse the situation. In this section, the situation in China is analysed and a proposal is explained for a framework of Payment for AGroEcosystem Services (**PAGES**) policy from the perspective of the political ecology of ecosystem services under the school of agroecology.

3.1 The Necessities of PAGES in China: From the Perspective of Political Ecology

3.1.1 A Political-Economic-Analysis of Farmer Sharing Income Declining in China

An increasing number of scientists have realized that producers in the agricultural sector share less and less among the full agricultural food chain's economy (Biovision Foundation for Ecological Development and IPES-Food 2020). This is not only because agriculture has become a steadily declining proportion of the national economy under the modern industrial division, but also because it is deeply embedded in the underestimation of the value of food that farmers provide (represented by Provisioning Services in Ecosystem Service concept). The value of its environmental services is not recognized and calculated at all in the current marketing system (Turner et al. 1988; Costanza et al. 1997). It will be useless to discuss the resuscitation of agriculture if we do not touch the issue of revaluing provisioning services that farmers generate. The monetary value for environmental services of agroecosystem needs to be recognized and it will be vital to increase public and policymakers' awareness of agriculture and food system, allowing for a decent solution of farmers' low income problems (de Groot et al. 2012; Voglhuber-Slavinsky et al. 2023).

With the world's second largest population, food security has always been a core issue for governance, policymaking, and scientific/technological issues for the Chinese government (Liu and Costanza 2010). However, since the newly founded country regime in 1949, the state-intervened economic system has greatly affected the balance between rural agriculture and urban industries, including agricultural production organization and relations between rural and urban; this in turn has had a huge impact on the status of peasants economically and politically (Chen et al. 2018).

China adopted the rural-urban divide policy both economically and politically when the People's Republic of China was founded in 1949 (Zhu 1999). Modelled on the former Soviet Union's 'the unequal exchange of different economic components under the condition of market isolation at home and abroad (Sau 1976), rural China and agriculture provided the primary accumulation in the economy for urban growth and its industrialization (Yang 1999), which was denoted as 'Extractive Institutions' (Acemoglu and Robinson 2012). Starting with the state practices of unified purchase and marketing of agricultural products in 1953, the following was established:

- Food Supply: Food from rural agriculture was supplied to urban community residents usually working in industries.
- Material Supply: Rural production agriculture offered raw products (such as food, tobacco, and fur) to urban processing industries at a cheap price and those processing industries sold the products back to the rural area at a high price.
- Marketing Capacity: Peasant purchased fertilizer, materials, machines, and daily needs, including processed food from urban industries at a high price.
- Factors Contribution: The rural area offers land and capital (via agricultural tax, industrial and agricultural products, and net capital outflow) to urban people and industries.
- Foreign Exchange Sink: Rural produces export-oriented products for trade but the returns go to urban industries for economic development.

In short, economically, money flowed from rural agriculture to subsidize the urban area, making rural peasants poorer, and ecologically, agricultural land was damaged and the rural living environment degraded. The urban industries grew and prospered at the social and economic expense of peasants. Peasants lost human dignity, and urban residents grew richer. This produced a huge divide between rural and urban citizens. According to Wen Tiejun in his famous book recently *The Eighth Crises: Lessons from China, 1949–2009* (Wen 2014), rural areas and agriculture became the land of the historical eighth economic crises, mostly due to political policy and actions. What followed was the abandonment of rural villages and degradation of agriculture; subsequently, this has hampered the healthy economic evolution of China. Since 2004, the Chinese government has changed priorities to rural and agriculture by issuing a document called 'The First Priority File' to remediate the situation, named as 'the Feedback Institution', aimed at 'to promote agriculture through industry, to lead the countryside through cities', the iconic events including 'the Construction of new socialist countryside' since 2005 and 'Revitalizing Countryside' since 2018 (Liu et al. 2018; Liu and Li 2017; Chen et al. 2018).

However, the succeeding 15 years of 'The First Priority File' (2004–2018) has not and will not reverse the situation using a need patchwork of repairs. The economic, ecological, and social predicaments in rural areas and within the agricultural sector are still expanding, with an increasingly enlarging gap in income between rural and urban citizens, with rural areas having substandard facilities of educational and medical services (Wang et al. 2012).

Given such a desperate situation of agriculture and rural areas in China, not only a certain degree of governmental intervention is necessary but also finding a proper entry point of policy will be extremely imperative. Academically, this is worth forcing ecologists, economists, and policymakers into serious discussions. There remains a huge gap between urban and rural, agriculture and industries, peasant and urban resident, and producer and consumer. It requires a serious dialogue between economics and ecology under the concept of ecosystem services and paying for its true value to society. Ideologically, there is an urgent need for a framework of political ecology to take the place of the prevailing but obsolete framework of political economics (Wang 2019; Jarosz 2012; Gruening 1951).

3.1.2 A Political-Ecological Analysis of Farmer Sharing Income Declining in China

With more than 1.4 billion people, China ranks as the second largest nation, but geographically and demographically, China is one of the poorest resource countries in the world, with per capita freshwater only one-fourth of world average, per capita arable land only one-fifth of the world average, and so on. However, since the reforming and opening policy at the beginning of the 1980s, China has set up the core development strategy on the economy by mobilizing every factor of natural and social resources. As a result, China has kept the average gross domestic production (GDP) increasing at a rate of nearly 10%, and by 2011, its GDP has been listed as the second of the world, next to the GDP of the USA. However, it should be noted that this has occurred with a huge ecological and social cost (Liu and Diamond 2005; Fang and Kiang 2006; Vandermeer 2011; Duru and Thérond 2015), because no country could sustain its economy rapid growth without damaging its environment and resources (MacBean 2007). Besides this ecological degradation, the social cost of the rapid growth characterized by huge economic disparities and knowledge gaps between rural and urban, agriculture and industries, poor and rich, and east-south sea coastal and west-north mountains is dramatically increasing.

Given such severe ecological and social sacrifices, the economic development paradigm in China should be transformed to sustainable development, in which the capital-oriented ideology of political economics should be transitioned to a society-oriented ideology of political ecology. This means that ecological sciences must play a dominant role over economics politically. The Ecological Civilization strategy was put forward by China's central government and was the right thing to undertake (Xiao and Zhao 2017; Liu et al. 2018; Liu 2010).

As the primary economic sector, agriculture is positioned between natural ecosystems and social systems; therefore, it encounters dual vulnerabilities in China. On the one hand, farmers get poorer, leading to the rural economic dilemma; on the other hand, rural China suffers much more severe environmental deterioration than do urban regions. This is the impact under a GDP growth-centred paradigm (Wang et al. 2012). A GDP priority farming system has forced farmers to use increasing amounts of chemical inputs: (1) China utilized 5900 million tons of fertilizer on arable land in 2018, which was 35.33% of the total fertilizer used in the world with the per capita fertilizer input of 258 kg.hm^{-2}, more than three times the world average; (2) China used 180.69 million tons of pesticides on arable land in 2018, which was 51.6% of the total pesticides used in the world, with the per capita pesticides input of 10.6 kg.hm^{-2}, more than 4.7 times that used in the USA (Wang 2019). This level of chemical use has led to arable land degradation and contamination, accompanied by universal loss of agrobiodiversity. The result is the dual crises of food security and safety in the present Mainland China. Hence, an agroecology-guided political ecology framework is more imperative to facilitate its transition to sustainability. The concept of ES provides an agro-eco-political foundation to fill the economic and knowledge gaps. Integrating ES concepts with agroecology concepts will provide a very good conceptual framework to steer the agroecological transitions (Dendoncker et al. 2018; Gliessman 2009). The integrating knowledge of agroecology and politics is vital to rescue China's desperate agriculture and rural dilemmas (De Molina 2013). Designing and implementing an appropriate PAGES will be the key for encouraging the provision of ES by farmers in China on the way to rural China's revitalization (Liu and Li 2017); this is the viable alternative to the policy and technology invention under the old framework of political economics in which the resources are destroyed (Wang 2019).

3.2 *PAGES: A Starting Point for Facilitating China's Agroecological Transitions*

Integration of thinking that combines natural and social sciences with humanities will be the key for future humankind survival (Barrett and Barrett 2001). But how can scientists from various disciplines understand China's agricultural and rural ecosystems and then unite to convey their ideas and solutions? This message must be conveyed to each stakeholder involved to maintain and sustain the provision of agroecosystem services and realize agricultural multifunctionalities (Garbach et al. 2016; Daelemans et al. 2023). Government may play a key role to explore the financial leverage in creating incentives for supplying the agroecosystem services (AGES) by transferring commodity-based subsidies to policy-based incentives in agriculture (Liu et al. 2008). Among which, PAGES or in China's tongue, the Ecological Compensation (EC) might be evaluated to see if it is an effective one (Shang et al. 2018; Fen et al. 2011; Liu et al. 2014).

Worldwide, payment for ecosystem services (PES) has been applied to transfer economic incentives to ecological protection. In the past two decades, there have been many proposed models (Engel et al. 2008; Banerjee et al. 2013). PAGES, driving from ES and PES concepts by using the combination of ecology and economics methodologies to produce economic leverage for protecting agro-environment via protecting farmers is proposed worldwide (Food and Agriculture Organization 2007; Ramírez-Suárez et al. 2024). Farmers have key roles in both maintaining and degrading ES via farming decisions (Power 2010; Zhang et al. 2007; Yaregal and Sime 2024). The significances of designing and implementing PAGES for China's agroecological transitions are:

- To embody the understanding that agroecosystems (via an economic industry named agriculture) in China are the fundamental life support system for the 1.4 billion people. By designing and implementing a PAGES (nationally and locally), the whole population of China will gradually gain a genuine understanding of why and how we do agriculture, i.e. utilizing plant's photosynthesis to assimilate the unlimited and clean solar energy for provision of the balanced ES to make people sustainably happy and healthy (Caldwell 1996; Wang 2005). As such, we expect to eliminate the obstacles for the existing thousand years of 'The Triple-Issues of Agriculture' in China.

- To establish the farmer's central role in the balance of ES provisions from agro-ecosystems. Essentially, AGES originates from integrative decision-making and labour by individual farmers; it is an integrative outcome of a farmer exploring his or her own natural, financial, physical, and social capitals (Smukler et al. 2012; Costanza et al. 2014), which is moulded by international and national agricultural and environmental policy in a great degree (Food and Agriculture Organization 2007). The latter thus is blamed for the externalities that agriculture creates and interests that farmer erodes, and the imbalanced ES that agroecosystems produce. Designing and implementing PAGES in China is considered as the main countermeasure to the market and governance dual failures, resulting in universal agroecosystem degradation, and will reinvent a way to internalize the externalities and sustainably provide agroecosystem services, not limited to PS but also its environmental services with synergies or trade-offs when farmers acquire motivation to follow the good agricultural practice (GAP) by such economic incentives (Power 2010).
- To help to monetize farmers' Natural Capital as the way to increase income of poor farmers living in ecologically fragile environments, which will be able to really create incentives for their performance in GAP and thus to produce PS and environment services sustainably.

3.3 Designing a Framework of PAGES Mechanism for China

3.3.1 From PES to PAGES: A Brief History

Worldwide, there are several PES programs that have been implemented; the most famous one was 'the Catskill/Delaware Watershed Agricultural program' in New York city. It consisted of a payment budget with 1–1.5 billion dollars for farmers instead of water cleaning engineering budget of 6.0–8.0 billion dollars (Rosa et al. 2004; Naeem et al. 2015). Since then, several Conservation Reserve Programs have been designed to pay landowners to retire millions of acres of fragile farmland. In the USA, it was called 'Grass them Down'; in England it was given the name of Agri-Environmental Scheme (AES). These institutional payments which recognize the value of agricultural multifunctionality, evolved into the Common Agricultural Policy (CAP) in the European Union (Dobbs and Pretty 2004). TEEB (The Economics of Ecosystems and Biodiversity) was established in the European Union in 2010 under the prevailing concept of ES internationally (Brink 2011; Kumar 2010; Ring et al. 2010; Du et al. 2016). Those programs are associated with agriculture and food systems in Europe and by linking land use transformation, landowners (not limited to farmers) were compensated for providing ecosystem services by managing, conserving, and enhancing the natural environment.

In the developing world, South American countries such as Mexico, Costa Rica, Brazil, and South Asian countries such as China and Indonesia are leading the PES programs, but the results vary. Those countries following the development world used ES concepts and methods to pay directly to the villagers to encourage them to protect and maintain their living and producing environments. For example, there have been two types of PES-like programs implemented in China, 'Grain for Blue' since 1998 (Song et al. 2010) and 'Grain for Green (also called 'Sloping Land Conversion Program (SLCP)' since 1999 (Uchida et al. 2009). Those programs have been positively evaluated in most research. 'Grain for Green' will be discussed in detail in the next section as the case for informing our model of PAGES. However, some programs initiated in South American countries have not been as successful, for example, 'the ScolelTé' program in Mexico was designed to help the indigenous farmers to participate in international carbon offset plans (Tipper 2002.), and 'the Pagesx' program in Costa Rica during 1997–2002 (Rosa et al. 2004.) was criticized as being low cost but low efficiency and perhaps even wasteful financially(Food and Agriculture Organization 2007).

When one considers the new understanding of agriculture linking food production to the goal of healthy and happy people, the current PES examples above were not directly linked to real 'agriculture' (Wang 2005; Wang et al. 2012; Caldwell and Wang 2020). They were PES policy like 'Payment for Forest Service Program' on arable land. PAGES proposed in this paper aims at PAGES designing and implementing for real food-producing oriented agriculture and farmers while maintaining environmental services. In this regard, policy for managing and realizing agroecosystem services is an untold story, even though it has been touched and implemented in some regions of the developed countries. For example, the Alternative Land Use Services (ALUS) programs designed for farmers in some provinces in Canada have significant incentives on GAP since 2010, but the conceptual framework of PAGES in agriculture today needs to be transitioned in a big step, changing from protecting the environment to protecting producer, and then in turn to pose better feedback on agriculture landscapes, and most important to final agricultural sustainability. This means the policy should emphasize not only soil erosion and land degradation but also the promotion of the whole spectrum of agroecosystem services; in other words, provisioning, regulating, supporting, and cultural services are equally important for human welfare provided by agricultural landscapes. The regulating, supporting, and cultural services of agroecosystems are indispensable components for achieving agricultural sustainability and offering quality of life not only for rural people but also for urban residents. As addressed above, China is a country suffering from the urban and rural divide policy which has led to a poor rural economy, degraded environment, and society dilemma. It will be an appropriate observation case to assess, design, and realize PAGES in China as a significant example for the rest of world.

3.3.2 Options for Capitalizing Natural Assets of Agroecosystem

Generally, there are four options to capitalize natural assets of agroecosystem, marketing return, subsidy (governmental), reward for the ecosystem goods or provisioning services, and penalty for the bad (governmental) and PAGES (direct payment), each depending on the concrete situations (Fig. 3.13).

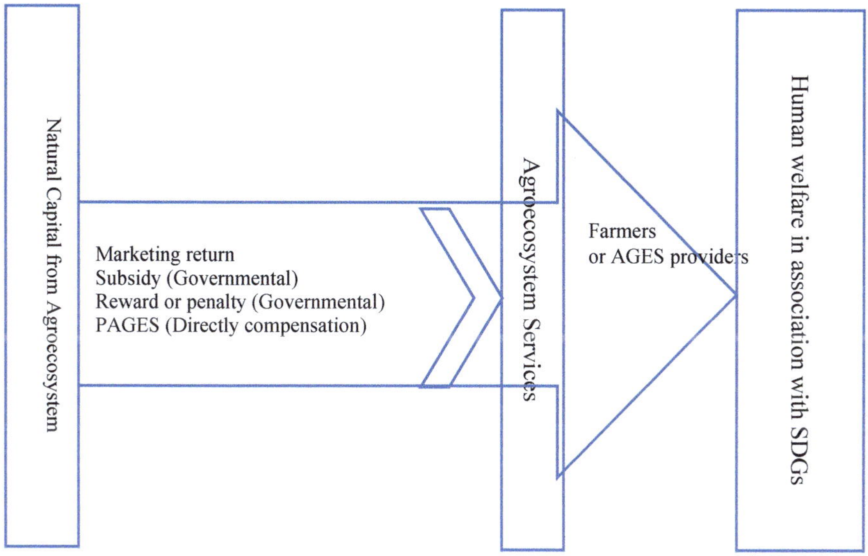

Fig. 3.13 Ways for capitalizing the natural asset of agroecosystem

Among them, PES is considered the most effective and efficient due to a direct linkage between payment and environmental service performance and is different from other mandatory mechanisms. The payment by beneficial sides represented usually by government or the third parties will guarantee no interest conflicts (Naeem et al. 2015).

According to cases described by FAO (2007) and other newly published papers(e.g. Novikova et al. 2017; Shang et al. 2018; Wang et al. 2016; Bohlen et al. 2009)， PAGES can only be appropriately established based on proper identification and assessment of agroecosystem services. By integrating the knowledge, we herein propose a general process for establishing PAGES for China.

3.3.3 China's Own Experience for Designing PAGES: Grain for Green

China is one of the first countries to initiate a PES program and address Ecological Compensation for remediating severe environmental degradation, a PAGES relevant program called the Sloping Land Conversion Program, also famous for the name 'The Grain for Green was initiated in 1999 (Liu et al. 2008) and was aimed at preventing fragile agrarian land (sloping land) from further eroding and also conserving the water cycle by paying directly the farmers to plant trees on the sloping land rather than growing crop as they did in the past. The initial objective was to plant trees on 14.67 million ha with a budget in the first Five-Year Plan period (2000–2005) of CNY 900 billion ($132 billion) and the second of CNY1,530 billion ($225 billion). 'Grain for Green' has been considered a success with few criticisms (Xu et al. 2014; Bennett 2008; Uchida et al. 2009). However, as mentioned above,

Grain for Green is not a true PAGES according to our definition of AGES which says it should be linked to food security and food safety. It was rather a continuity of the Natural Forest Conservation Program (NFCP) which was initiated in the 1990s by China, focusing on provision of ES of forestry rather than on AGES.

3.3.4 Learn from Canadian Case: Alternate Land Use Services (ALUS)

In general, worldwide since 2008, there has been an increasing amount of attention paid to the preservation of natural capital and ecosystem services. According to Ruggeri (2009), Canada has spent significant attention to the concept of natural capital and ES by linking the policymaking processes to the national economic chain. Regarding the general ES, three incentive-based policies are used in Canada: Purchased Property Rights (PPRs), Tradeable Development Rights (TDRs), and Development Impact Fees (DIFs). Some provincial incentive-based policies have been particularly effective, meaning that compensation directly reaches landowners to secure the NC and ES, which has extended to AGES since 2002. For example, a PAGES program named ALUS was implemented in several provinces of Canada, including Prince Edward Island, Manitoba, Ontario, Alberta, Saskatchewan, and British Columbia.

Alternative Land Use Services (ALUS) is a community-led, farmer-delivered program that provides a financial incentive for producing valuable ecosystem services on Canadian farmland; it supports stewardship activities on agricultural lands (France and Campbell 2015). Farmers are paid to change the farming methods from conventional (productive system harms natural habitat) to conservational farming. This program was first designed and implemented in 2000 in Manitoba for Keystone Agriculture Producers, then gradually extended to Prince Edward Island, Ontario, Alberta, Saskatchewan, and British Columbia during 2002–2011, empowering those provinces to aggregate funding for PS and Environmental Services. The money sources vary from conservation funding, mitigation, offsets, philanthropic supports, and nontraditional sources. It is voluntary, incentive-based, non-regulatory, local government and agricultural agencies delivered, community-led program, recognizing the important roles farmers and ranchers play as stewards of the land. It empowers them to deliver nature's benefit on the working landscape. It is a model for producer-led sustainability of agriculture and food system. In British Columbia, the NGO foundations were built to accumulate the finance from various bodies, including the private landowners themselves, the latter could refund the money if they conduct GAP. ALUS is organized on recognizing the farmer's key role in creating a healthy environment to embrace wildlife while producing sufficient food to support human life. Thus, we consider ALUS is a good model in designing and implementing PAGES for steering China's agroecological transitions.

3.3.5 Designing a PAGES for Steering China's Agroecological Transition

MEA (2003) classified the ES into four categories (provisioning, supporting, regulating, and cultural services), but particular cases vary with institutions and therefore have implications for evaluation (Mäler et al. 2008). In China, agroecosystem services relate mainly to the socioeconomic-political dimensions. A PAGES framework proposed here is as follows:

Linguistically, PES is a means to apply economic methods to aggregate financial and institutional stakeholders of natural resources for using natural resources sustainably. The components of a PES usually include the assessing methodologies, compensating criteria, and compensating routes (Food and Agriculture Organization 2007; Banerjee et al. 2013; Engel et al. 2008; Daily et al. 2009). Considering the experience and lessons from previous PES programs, the basic components are presented as an interactive cycle in Fig. 3.14 (Yin et al. 2013; Jack et al. 2008).

- **Identifying AGES:** AGES identification is a primary step of a PAGES strategy, an appropriate identification AGES targets to synergize and trade-off the provisioning services (market value) and environmental services (non-market value) of the agroecosystem. Costanza et al. (1997) identified and evaluated 17 types of ES in 11 global terrestrial ecosystem; among them, only three kinds of AGES (pollination, biological control, and food production) were recognized due to insufficient data which has caused further evaluation (Fisher et al. 2009; Costanza et al. 2014). There are more AGES provisions that need to be quantified, such as carbon sequestration, biodiversity-wildlife habitat, water filtration, pollinator habitat, erosion control, species-at-risk habitat, flood control, drought mitigation, climate adaptation, and community resilience. It has also been argued that both a macro-view and a local view on AGES identification should be integrated in any model (Simon-Rojo 2023).

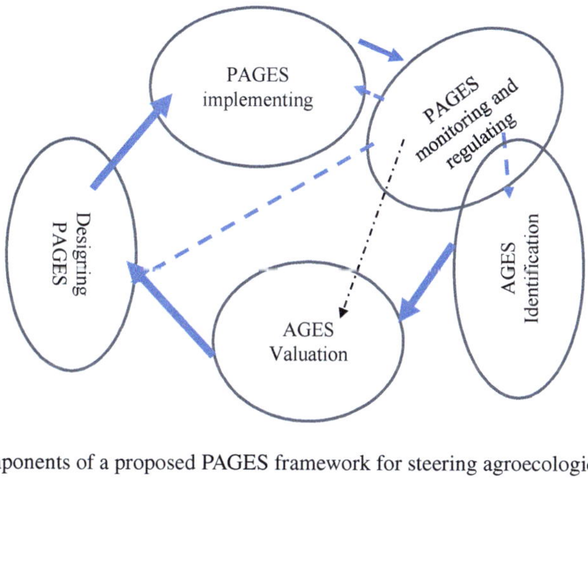

Fig. 3.14 Components of a proposed PAGES framework for steering agroecological transitions

- ***Valuing AGES:*** The concept and classification of ES clarified the importance of ES to human beings (MEA 2003). Valuing ES further makes clear what degree the natural capital will be recognized and calculated. Since Costanza et al. (1997)'s initiative on ES valuation in global ecosystems (Costanza et al. 1997), there have been numerous case studies on valuing ES worldwide. Three debatable issues remain, one is how to assign proportionally the regional and local ES values among the global ES values, given the global ecosystem could not simply be divided into regional and local ecosystems; another is how to integrate the various landscapes of ES (Jacobs et al. 2018). Another one is to seek an integrative method to evaluate the bundles of ES, or the total value of ES (Raudsepp-Hearne et al. 2010). Also, among the market-valued PS, energy value should be added, naming it Combined Food and Energy (CEF) is appropriated (Porter et al. 2009). In AGES regime, the qualification and balance of the ES and disservice value is also challenging (Zhang et al. 2007; Sabatier et al. 2013; Costanza et al. 2014). Figure 3.15 sketches the strategies of valuing AGES in order to design an appropriate PAGES in local scale, in which agroecosystem intersects with natural ecosystem and social-economic-political systems. AGES consists of PS valued by market pricing process and socializing investment and environmental services valued by public benefit valuing process and socializing investment. The socializing part of valuing AGES could be done by surveys of a-willing-to-pay and to-accept methods (Farley et al. 2010).

In China's situation, PS of agroecosystems with characteristics of both private and public goods should be appropriately valued by both market and society, and

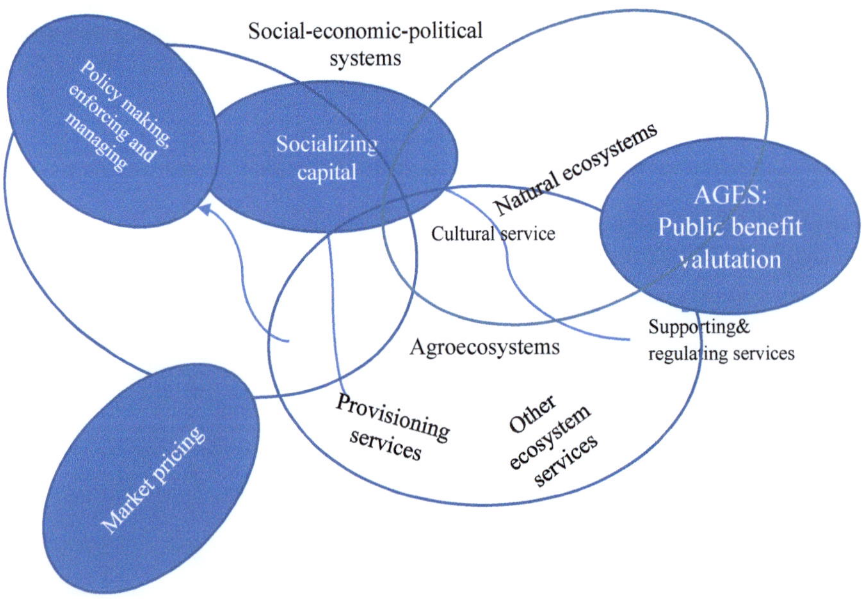

Fig. 3.15 The strategies of valuing AGES

environmental services of agroecosystem as completely public goods will be compensated economically to farmers by the government, society, and newly created market after appropriate assessment is made.

In terms of value methodologies, since Costanza et al. (1997) evaluated global ecosystem services (Costanza et al. 1997), there have been several case studies copying their approach and agroecosystem services sometimes are assessed. However, due to the distinct differences between ecosystems and agroecosystems, the assessing methodologies remain controversial. For instance, Xie et al. (2008) employed the Delphi method to show that agroecosystem services are heavily underestimated when following the methodologies that Costanza et al. (1997) used, and ecosystem services in developing countries are overestimated according to their economic standards, and some services are not appropriately assessed due to data set deficiency (Xie et al. 2008). Therefore, in this regard, there is a need to establish regional assessment methodologies for agroecosystem services particularly when this assessment link to accurate financial offsets. Theoretically, willing to pay and willing to be compensated are equally important and could be an ideal method to calculate the agroecosystem services.

- *Designing a PAGES Policy:* The goal of designing a PES policy is for efficient ES provisions in the long run. There is an increasing literature on ES concepts and its evaluation in the last two decades after Gretchen C. Daily (1997) 's ES proposal and Costanza et al.'s ES evaluation (Satama-Bermeo et al. 2024), but sound PES policy designing is still in process even though there have been numerous PES projects initiated worldwide. Most of them are not cost-benefit efficient except for a few cases like China's 'Grain for Green' program (Jack et al. 2008). The real agroecosystem services-based payment designing is scarce (Power 2010). Therefore, designing a PAGES for China's urgent agroecological transition will set up an example for most developing countries whose agricultural GDP still occupies more than 20% implying that they are still a natural resource dependent economy (Dasgupta 2007).

  There has been a consensus that including economy of nature into human economic system is a necessary step when facing environmental problems (Barbier 2011; Daily et al. 2000); in this regard, eco-compensation is one means to capitalize the natural asset in the name of natural capital (Ouyang et al. 2016). Unless this is compensated, ES degradation will be an unavoidable fate until it reaches the level of a threat to human survival (Costanza 2006). One could imagine the disastrous consequence of AGES dysfunction and disappearance. In this respect, PAGES should be designed imperatively. By integrating the relevant PAGES literature to date, the following is a proposed general framework of PAGES for China's agroecological transitions (Fig. 3.16).

  As shown in Fig. 3.16, a practical PAGES program requires several appropriate incentives to encourage farmers to join the program to apply GAP to enable a functioning AGES. The winnowing process starts to screen the eligible participants after a certain agroecosystem is chosen for the program (Weinberg and Claassen 2006; Polasky 2008). This depends on the producers' farm type (the

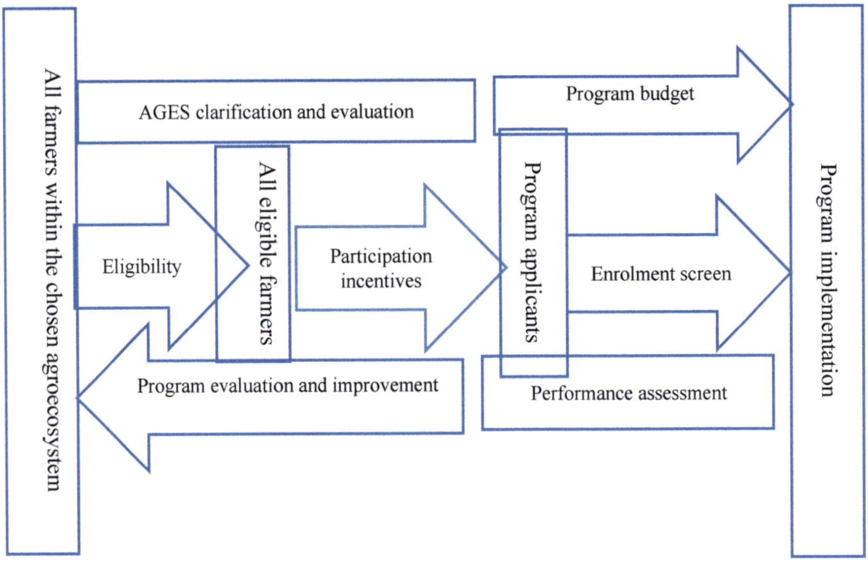

Fig. 3.16 A general framework of PAGES for steering China's agroecological transitions

agroecosystem structure spatially and temporally). The payment level that produces incentives is determined by the values of agroecosystem services that are evaluated. Thereafter the budget is assigned accordingly and the payments from AGES are released to the farmers based on what they create and preserve.

There are lessons and experiences from the Canadian ALUS program since 2008; the financial funding is the key to implement the PAGES because of its public good property, and the funding should be led by the government and complemented by marketing power and community empowerment, participating by farmers.

To sustain the program, the cost-effectiveness and program performance ought to be paid, evaluated, and then feedback for next round of possible revisions and improvement.

- *Implementing a PAGES Policy:* Once a PES policy is designed, the route of implementation will be critical for its cost-effectiveness (Food and Agriculture Organization 2007; Uchida et al. 2005). The route setup and time line are two keys in implementing stages. China's Grain for Green project was dominated by government (Yang et al. 2013), the ALUS project in Canada since 2008 was another successful project which was initiated by community in association with environmental conservation and agriculture and food and participated by farmer and ranchers voluntarily. It is advisable to draw some experiences and lessons for future more PAGES implementation in China (Fontana et al. 2013; Bennett 2018; Mouchet et al. 2017).

Evaluating the PAGES policy: Evaluating a PAGES policy is a final stage of a PAGES policy loop that records the key notes while implementing and evaluating the outcome of PAGES policy after implemented for the sake of improving the policy for the next round implementation and review the policy (Geneletti 2013; Dressler et al. 2015; Lerouge et al. 2016; Clapp et al. 2018; Kain et al. 2016; Barral et al. 2015; Dale and Polasky 2007).

In sum, this section aims at taking China's situation as a case showing a regional rather than a global agroecosystem, by touching its depth and breadth, to design the PAGES framework with implications for policy and applications. The process is as follows: (1) analyse the urgent state of China's agricultural and rural dilemmas attaching it to an agroecological transition road map through the view of political ecology, (2) review the current research state and trend of PAGES worldwide by examining the existing PAGES programs worldwide particularly in Canada and China, and (3) explore a framework of PAGES for steering China's agroecological transitions (with a case of AGES evaluation of Fujian Province, China being drafting). This case helps in the exploration of several knowledge gaps in practising PAGES worldwide.

References

Acemoglu, D., and J. A. Robinson. 2012. *Why nations fail: the origins of power, prosperity, and poverty*. New York: Crown Business. book review.

Babin, Nicholas. 2015. The coffee crisis, fair trade, and agroecological transformation: impacts on land-use change in Costa Rica. *Agroecology and Sustainable Food Systems* 39 (1): 99–129.

Barbier, E. 2011. *Capitalizing on nature: ecosystems as natural assets*. Cambridge: Cambridge University Press.

Barral, M. P., J. M. Rey Benayas, P. Meli, and N. O. Maceira. 2015. Quantifying the impacts of ecological restoration on biodiversity and ecosystem services in agroecosystems: a global meta-analysis. *Agriculture, Ecosystems and Environment* 202:223–231.

Barrett, G. W., and T. L. Barrett. 2001. *Holistic science: the evolution of the Georgia Institute of Ecology (1940–2000)*. New York: Taylor & Francis.

Banerjee, Onil, Neville D. Crossman, and Rudolf S. de Groot. 2013. Ecological processes, functions and ecosystem services: inextricable linkages between wetlands and agricultural systems. In *Ecosystem services in agricultural and urban landscapes*, ed. Steve Wratten, Harpinder Sandhu, Ross Cullen, and Robert Costanza, 1st ed., 16–26. Wiley.

Bennett, M. T. 2008. China's sloping land conversion program: institutional innovation or business as usual? *Ecological Economics* 65 (4): 699–711.

Bennett, Michael T., Yazhen Gong, and Riccardo Scarp. 2018. Hungry birds and angry farmers: using choice experiments to assess "eco-compensation" for coastal wetlands protection in China. *Ecological Economics* 154:71–87.

Biovision Foundation for Ecological Development & IPES-Food. 2020. Money flows: what is holding back investment in agroecological research for Africa? *Biovision Foundation for Ecological Development & International Panel of Experts on Sustainable Food Systems*. www.biovision.ch and www.ipes-food.org. Accessed February 11, 2025.

Bohlen, P., S. Lynch, L. Shabman, M. Clark, S. Shukla, and H. Swain. 2009. Paying for environmental services from agricultural lands: an example from the northern Everglades. *Frontiers in Ecology and the Environment* 7 (1): 46–55.

Brink, P. T. 2011. *The economics of ecosystems and biodiversity in national and international policy making*. London: Earthscan.

Bureau d'analyse sociétale d'intérêt collectif (BASIC). 2024. *The Grounds for Sharing: A study of value distribution in the coffee industry*. https://www.solidaridadnetwork.org/publications/the-grounds-for-sharing-a-study-of-value-distribution-in-the-coffee-industry. Accessed Dec.2, 2025.

Caldwell, C.D. 1996. *Agricultural ecosystems*. Unpublished course manual, Dalhousie University, Truro, Canada.

Caldwell CD, Wang SL (Eds) (2020) Introduction to agroecology. Springer, Singapore

Castro, R., Tattenbach, F., Olson, N., and Gamez, L. 1997. The Costa Rican experience with market instruments to mitigate climate change and conserve biodiversity. In *The Global Conference on Knowledge for Development in the Information Age*. Toronto, Canada.

Chen, C., R. LeGates, M. Zhao, and C. Fang. 2018. The changing rural-urban divide in China's megacities. *Cities* 81:81–90.

Clapp, C. M., J. Freeland, K. Ihlanfeldt, and K. Willardsen. 2018. The fiscal. Impacts of alternative land uses: an empirical investigation of cost of community services studies. *Public Finance Review* 46 (5): 850–878.

Conway, G. R. 1985. Agroecosystem analysis. *Agricultural Administration* 20 (31): 55.

Conway, Gordon R. 1986. *Agroecosystem analysis for research and development*. Bangkok: Winrock International.

Conway, Gordon R. 1987. The properties of agroecosystems. *Agricultural Systems* 24 (2): 95–117.

Costanza, R. 2006. Ecosystems without commodifying them. *Nature* 443 (7113): 749.

Costanza, R., R. d'Arge, R. de Groot, S. Farber, M. Grasso, B. Hannon, K. Limburg, S. Naeem, et al. 1997. The value of the world's ecosystem services and natural capital. *Nature* 387:253–260.

Costanza, R., R. de Groot, P. Sutton, S. van der Ploeg, S. J. Anderson, I. Kubiszewski, S. Farber, and R. K. Turner. 2014. Changes in the global value of ecosystem services. *Global Environmental Change* 26:152–158.

Daelemans, Robin, Eva Hulsmans, Lysander Fockaert, Liesbet Vranken, Luc de Bruyn, and Olivier Honnay. 2023. Agroecosystem multifunctionality of apple orchards in relation to agricultural management and landscape context. *Ecological Indicators* 154:110496.

Daily, Gretchen C., ed. 1997. *Natures services: societal dependence on natural ecosystems*. Washington, DC: Island Press.

Daily, Gretchen C., Susan Alexander, Paul R. Ehrlich, Larry Goulder, Jane Lubchenco, Pamela A. Matson, Harold A. Mooney, Sandra Postel, Stephen H. Schneider, David Tilman, and George M. Woodwell. 1997. Ecosystem services: benefits supplied to human societies by natural ecosystems. *Issues in Ecology* 1 (2): 1–18.

Daily, Gretchen C., T. Söderqvist, S. Aniyar, K. Arrow, P. Dasgupta, P. R. Ehrlich, and B. Walker. 2000. The value of nature and the nature of value. *Science* 289 (5478): 395–396.

Daily, Gretchen C., Stephen Polasky, Joshua Goldstein, Peter M. Kareiva, Harold A. Mooney, Liba Pejchar, Taylor H. Ricketts, James Salzman, and Robert Shallenberger. 2009. *Frontiers in Ecology and the Environment* 7 (1): 21–28. https://doi.org/10.1890/080025.

Dale, V. H., and S. Polasky. 2007. Measures of the effects of agricultural practices on ecosystem services. *Ecological Economics* 64 (2): 286–296.

Dasgupta, P. 2007. Nature and the economy. *Journal of Applied Ecology* 44 (3): 475–487.

de Groot, R., L. Brander, S. van der Ploeg, R. Costanza, F. Bernard, L. Braat, et al. 2012. Global estimates of the value of ecosystems and their services in monetary units. *Ecosystem Services* 1 (1): 50–61.

De Molina, M. 2013. Agroecology and politics: how to get sustainability? About the necessity for a political agroecology. *Agroecology and Sustainable Food Systems* 37 (1): 45–59.

Dendoncker, N., F. Boeraeve, E. Crouzat, M. Dufrêne, A. König, and C. Barnaud. 2018. How can integrated valuation of ecosystem services help understanding and steering agroecological transitions? *Ecology and Society* 23 (1): 12. https://doi.org/10.5751/ES-09843-230112.

Dobbs, T. L., and J. N. Pretty. 2004. Agri-environmental stewardship schemes and multifunctionality. *Review of Agricultural Economics* 26 (2): 220–237.

Dressler, Wolfram, David Wilson, Jessica Clendenning, Rob Cramb, Sango Mahanty, Rodel Lasco, Rodney Keenan, Phuc To, and Dixon Gevana. 2015. Examining how long fallow swidden systems impact upon livelihood and ecosystem services outcomes compared with alternative land-uses in the uplands of Southeast Asia. *Journal of Development Effectiveness* 7 (2): 210–229.

Du, L., J. Li, G. Liu, F. Zhang, J. Xu, and L. Hu. 2016. Progress in the researches on the economics of ecosystems and biodiversity (TEEB). *Biodiversity Science* 24 (6): 686–693.

Duru, M., and O. Thérond. 2015. Designing agroecological transitions: a review. *Agronomy for Sustainable Development* 35 (4): 1237–1257.

Ehrlich, P. R., and A. H. Ehrlich. 1981. *Extinction: the causes and the consequences of the disappearance of species*. New York: Random House & Ballantine.

Engel, S., S. Pagiola, and S. Wunder. 2008. Designing payments for environmental services in theory and practice: an overview of the issues. *Ecological Economics* 65:663–674.

Fang, J., and C. S. Kiang. 2006. China's environment: challenges and solutions. *Frontiers in Ecology and the Environment* 4 (7): 339.

Farley, Joshua, Abdon L. Schmitt, Juan P. Alvez, and Paola M. Rebola. 2010. The farmer's viewpoint: payment for ecosystem services and agroecologic pasture based dairy production. *Advances in Animal Biosciences* 1 (2): 490–491. https://doi.org/10.1017/S2040470010001093.

Fen, Li, Li Wenhua, Zhen Lin, Huang Heqing, Wei Yunjie, and Inoue Naomi. 2011. Estimating eco-compensation requirements for forest ecosystem conservation a case study in Hainan province, southern China. *Outlook Agriculture* 40 (1): 51–57.

Fontana, V., A. Radtke, Fedrigotti V. Bossi, U. Tappeiner, E. Tasser, S. Zerbe, and T. Buchholz. 2013. Comparing land-use alternatives: using the ecosystem services concept to define a multi-criteria decision analysis. *Ecological Economics* 93:128–136.

Food and Agriculture Organization. 2007. *The State of Food and Agriculture: Paying Farmers for Environmental Services*. FAO Agriculture Series, No. 38, Rome, 222 p.

France, R. L., and J. B. Campbell. 2015. Payment for agroecosystem services developmental case-history descriptions of Canada's grassroots 'ALUS' programs. *Research Journal of Agriculture and Environmental Management* 4:405–531.

Garbach, K., J. C. Milder, F. A. J. DeClerck, L. Driscoll, M. Montenegro, and B. Herren. 2016. Examining multi-functionality for crop yield and ecosystem services in five systems of agro-ecological intensification. *International Journal of Agricultural Sustainability* 15 (1): 11–28.

Geneletti, D. 2013. Assessing the impact of alternative land-use zoning policies on future ecosystem services. *Environmental Impact Assessment Review* 40:25–35.

Gliessman, Stephen R. 2004. Agroecology and agroecosystems. In *Agroecosystems analysis*, ed. D. Rickerl and C. Francis, 19–30. Madison, WI: American Society of Agronomy.

Gliessman, S. R. 2009. The framework for conversion. In *The conversion to sustainable agriculture: principles, processes, and practices*, ed. S. R. Gliessman and M. Rosemeyer, 3–16. Boca Raton, FL: CRC Press.

Gruening, E. 1951. The political ecology of Alaska. *The Scientific Monthly* 73 (6): 376–386.

Hong, Qian. 2012. *China: symbiotic rise*. Beijing: Intellectual Property Publishing House.

Jack, B. K., C. Kousky, and R. E. Sims Katharine. 2008. Designing payments for ecosystem services: lessons from previous experience with incentive-based mechanisms. *Proceedings of the National Academy of Sciences* 105 (28): 9465–9470.

Jacobs, Sander, Berta Martín-López, David N. Barton, Robert Dunford, Paula A. Harrison, Eszter Kelemen, Heli Saarikoski, Mette Termansen, Marina García-Llorente, Erik Gómez-Baggethun, Leena Kopperoinen, Sandra Luque, Ignacio Palomo, Joerg A. Priess, Graciela M. Rusch, Patrizia Tenerelli, Francis Turkelboom, Rolinde Demeyer, Jennifer Hauck, Hans Keune, and Ron Smith. 2018. The means determine the end—pursuing integrated valuation in practice. *Ecosystem Services* 29:515–528.

Jarosz, Lucy. 2012. Growing inequality: agricultural revolutions and the political ecology of rural development. *International Journal of Agricultural Sustainability* 10 (2): 192–199.

Kain, J., N. Larondelle, D. Haase, and A. Kaczorowska. 2016. Exploring local consequences of two land-use alternatives for the supply of urban ecosystem services in Stockholm year 2050. *Ecological Indicators* 70:615–629.

Kull, Christian A., Xavier Arnauld, Monica de Sartre, and Castro-Larranga. 2015. The political ecology of ecosystem services. *Geoforum* 61:122–134.

Kumar, P., ed. 2010. *The economics of ecosystems and biodiversity: ecological and economic foundations*. London: Earthscan.

Lerouge, F., K. Sannen, H. Gulinck, and L. Vranken. 2016. Revisiting production and ecosystem services on the farm scale for evaluating land use alternatives. *Environmental Science and Policy* 57:50–59.

Liu, J. 2010. China's road to sustainability. *Science* 328 (5974): 50. https://doi.org/10.1126/science.1186234.

Liu, S., and R. Costanza. 2010. Ecosystem services valuation in China. *Ecological Economics* 69 (7): 1387–1388.

Liu, J. G., and J. Diamond. 2005. China's environment in a globalizing world. *Nature* 435 (7046): 1179–1186.

Liu, Y., and Y. Li. 2017. Revitalize the world's countryside. *Nature* 548 (7667): 275.

Liu, J., S. Li, Z. Ouyang, C. Tam, and X. Chen. 2008. Ecological and socioeconomic effects of China's policies for ecosystem services. *Proceedings of the National Academy of Sciences of the United States of America* 105 (28): 9477–9482.

Liu, M. C., Y. Xiong, Z. Yuan, Q. W. Min, Y. H. Sun, and A. M. Fuller. 2014. Standards of ecological compensation for traditional eco-agriculture: taking rice-fish system in Hani Terrace as an example. *Journal of Mountain Science* 11 (4): 1049–1059.

Liu, Chen, Lily Chen, Robert M. Vanderbeck, Gill Valentine, Mei Zhang, Kristina Diprose, and Katie McQuaid. 2018. A Chinese route to sustainability: postsocialist transitions and the construction of ecological civilization. *Sustainable Development*. 1–8.

MacBean, A. 2007. China's environment: problems and policies. *World Economics* 30 (2): 292–307.

Mäler, Karl-Göran, Sara Aniyar, and Åsa Jansson. 2008. Accounting for ecosystem services as a way to understand the requirements for sustainable development. *Proceedings of the National Academy of Sciences of the United States of America* 105 (28): 9501–9506.

Matulis, B. S. 2014. The economic valuation of nature: a question of justice? *Ecological Economics* 104:155–157.

Millennium Ecosystem Assessment. 2003. *Ecosystem and human well-being: framework for assessment*. Washington, DC: Island Press. (chiefly edited by Alcamo J, Bennett EM (eds)).

Millennium Ecosystem Assessment. 2005a. *Ecosystems and human well-being: synthesis*. Washington, DC: Island Press.

Millennium Ecosystem Assessment. 2005b. *Ecosystems and human well-being: global assessment reports*. Washington, DC: Island Press.

Mouchet, M. A., C. Rega, R. Lasseur, D. Georges, M. Paracchini, J. Renaud, J. Stürck, C. J. E. Schulp, P. H. Verburg, P. J. Verkerk, and S. Lavorel. 2017. Ecosystem service supply by European landscapes under alternative land-use and environmental policies. *International Journal of Biodiversity Science, Ecosystem Services & Management* 13 (1): 342–354.

Naeem, S., J. Ingram, A. Varga, et al. 2015. Get the science right when paying for nature's services. *Science* 347 (6227): 1206–1207.

Novikova, A., L. Rocchi, and V. Vitunskienė. 2017. Assessing the benefit of the agroecosystem services: Lithuanian preferences using a latent class approach. *Land Use Policy* 68:277–286.

Ouyang, Z., H. Zheng, Y. Xiao, S. Polasky, J. Liu, W. Xu, Q. Wang, L. Zhang, E. Rao, L. Jiang, F. Lu, X. Wang, G. Yang, S. Gong, B. Wu, Y. Zeng, W. Yang, and G. Daily. 2016. Improvements in ecosystem services from investments in natural capital. *Science* 352 (6292): 1455–1459.

Polasky, S. 2008. What's nature done for you lately: measuring the value of ecosystem services. *Choices* 23 (2): 42–46.

Porter, John, Robert Costanza, Harpinder Sandhu, Lene Sigsgaard, and Steve Wratten. 2009. The value of producing food, energy, and ecosystem services within an agroecosystem. *AMBIO: A Journal of the Human Environment* 38 (4): 186–193.

Power, A. G. 2010. Ecosystem services and agriculture: tradeoffs and synergies. *Philosophical Transaction of Royal Society Biological Science* 365:2959–2971.

Ramírez-Suárez, Wendy Mercedes, Jorge Carlos López-Chouza, María de los Ángeles Flores-Acosta, Saray Sánchez-Cárdenas, and Pedro Lázaro Rodríguez-Morejón. 2024. Biodiversity and ecosystem services in agroecological systems. A review. *Pastosy Forrajes* 47:e02.

Raudsepp-Hearnea, C., G. D. Petersona, and E. M. Bennett. 2010. Ecosystem service bundles for analyzing tradeoffs in diverse landscapes. *Proceedings of the National Academy of Sciences of the United States of America* 107 (11): 5242–5247.

Ring, I., B. Hansjürgens, T. Elmqvist, H. Wittmer, and P. Sukhdev. 2010. 2010.Challenges in framing the economics of ecosystems and biodiversity: the TEEB initiative. *Current Opinion in Environment Sustainability* 2 (1): 15–26.

Rosa, H., S. Kandel, and L. Dimas. 2004. Compensation for environmental services and rural communities: lessons from the Americas. *International Forestry Review* 6 (2): 187–194.

Ruggeri, Joe. 2009. Government investment in natural capital. *Ecological Economics* 6 (8): 1723–1739.

Sabatier, Rodolphe, Katrin Meyer, Kerstin Wiegand, and Yann Clough. 2013. Non-linear effects of pesticide application on biodiversity-driven ecosystem services and disservices in a cacao agroecosystem: a modeling study. *Basic and Applied Ecology* 14:115–125.

Satama-Bermeo, Maritza, Max Rudolf, and Roland Olschewski. 2024. Agroecosystem services: from general assessment to policy implications. *People and Nature* 6 (3): 1060–1077.

Sau, R. 1976. The theory of unequal exchange, trade and imperialism. *Economic and Political Weekly* 11 (10): 399–404.

Shang, W., Y. Gong, Z. Wang, and M. J. Stewardson. 2018. Eco-compensation in China: theory, practices and suggestions for the future. *Journal of Environmental Management* 210:162–170.

Simoncini, R. 2009. Developing an integrated approach to enhance the delivering of environmental goods and services by agro-ecosystems. *Regional Environmental Change* 9 (3): 153–167.

Simon-Rojo, Marian. 2023. The role of ecosystem services in the design of agroecological transitions in Spain. *Ecosystem Services* 61:101531.

Smukler, S. M., S. M. Philpott, L. E. Jackson, A. Klein, F. DeClerck, L. Winowiecki, and C. A. Palm. 2012. Ecosystem services in agricultural landscapes. In *Integrating ecology and poverty reduction*, ed. Jane Carter Ingram, Fabrice DeClerck, and Cristina Rumbaitis del Rio, 1st ed., 17–51. Boston, MA: Springer.

Song, W., X. Deng, B. Liu, Z. Li, and G. Jin. 2010. Impacts of grain-for-green and grain-for-blue policies on valued ecosystem services in Shandong Province, China. *Advances in Meteorology* 2015:1–10.

Tansley, Arthur G. 1935. The use and abuse of vegetational concepts and terms. *Ecology* 16 (3): 284–307.

Tipper, R. 2002. Helping indigenous farmers participate in the international market for carbon services: the case of ScolelTé. In *Selling forest environmental services: market-based mechanisms for conservation and development*, ed. S. Pagiola, J. Bishop, and N. Landell-Mills, 223–234. London: Earthscan.

Turner, M. G., E. P. Odum, R. Costanza, and M. Thomas. 1988. Market and nonmarket values of the Georgia landscape. *Environmental Management* 12:209–217.

Uchida, E., S. Rozelle, and J. Xu. 2009. Conservation payments, liquidity constraints, and off-farm labor: impact of the grain-for-green program on rural households in China. *American Journal of Agricultural Economics* 91 (1): 70–86.

Uchida, Emi, Jintao Xu, and Scott Rozelle. 2005. Grain for green: cost-effectiveness and sustainability of China's conservation set-aside program. *Land Economics* 81 (2): 247–264.

Vandermeer, John. 2011. *The ecology of agroecosystems*, 315–316. Sudbury, MA: Jones and Bartlett Publishing.

Voglhuber-Slavinsky, A., N. Lemke, J. MacPherson, Ewa Dönitz, Mathias Olbrisch, Philipp Schöbel, Björn Moller, Enno Bahrs, and Katharina Helming. 2023. Valorization for biodiversity and ecosystem services in the agri-food value chain. *Environmental Management* 72:1163–1188. https://doi.org/10.1007/s00267-023-01860-7.

Wang, Songliang. 2005. Information technology: toward the way to the sustainable management of agroecosystem (in Chinese). *Agricultural Network Information* 8 (4–7): 12.

Wang, Songliang. 2007. Ecosystem service function assessment and management (in Chinese). In *Ecology*, ed. Lin Wenxiong, 1st ed., 277–284. Beijing: China Science Press.

Wang, Songliang. 2019. Concerted development of ecological agriculture along with community-supported agriculture to facilitate rural vitalization (in Chinese). *Chinese Journal of Eco-Agriculture* 27 (1): 212–217.

Wang, Songliang, Caldwell, C. D., Kilyanek, S. L., & Sziklay, J. 2012. Agroecology(in English). Beijing: China's Science Press.

Wang, H., Z. Dong, Y. Xu, and C. Ge. 2016. Eco-compensation for watershed services in China. *Water International* 41 (2): 271–289.

Weinberg, M., and R. Claassen. 2006. A multitude of design decisions influence conservation program performance. *Amber Waves* 4:2.

Wen, T. J. 2014. *Eight crises: lessons from China 1949–2009 (in Chinese)*. Beijing: Oriental Press.

Wood, Sylvia L. R., and Fabrice DeClerck. 2015. Ecosystems and human well-being in the sustainable development goals. *Frontiers in Ecology and the Environment* 13 (3): 123.

Xiao, L., and R. Zhao. 2017. China's new era of ecological civilization. *Science* 358 (6366): 1008–1009.

Xie, Gaodi, Lin Zhen, Chu-xia Lu, Yu Xiao, and Cao Chen. 2008. Expert knowledge based valuation method of ecosystem services in China (in Chinese). *Journal of Natural Resources* 23 (5): 911–919.

Xu, Z., M. T. Bennett, R. Tao, and J. Xu. 2014. China's sloping land conversion program four years on: current situation and pending issues. *International Forestry Review* 6 (4): 317–326.

Yang, D. T. 1999. Urban-biased policies and rising income inequality in China. *The American Economic Review* 89 (2): 306–310.

Yang, W., J. Liu, W. Liu, A. Viña, J. Luo, G. He, Z. Ouyang, and H. Zhang. 2013. Performance and prospects of payments for ecosystem services programs: evidence from China. *Journal of Environmental Management* 127:86–95.

Yaregal, Yihunie, and Getachew Sime. 2024. Traditional home garden agro-biodiversity dynamics, agro-ecosystem services, and management practices in smallholder farmers setting, South-Central Ethiopia. *Food and Energy Security* 13:e569. https://doi.org/10.1002/fes3.569.

Yin, Runsheng, Tianjun Liu, Shunbo Yao, and Minjuan Zhao. 2013. Designing and implementing payments for ecosystem services programs: lessons learned from China's cropland restoration experience. *Forest Policy and Economics* 35:66–72.

Zhang, W., T. H. Ricketts, C. Kremen, K. Carney, and S. M. Swinton. 2007. Ecosystem services and dis-services to agriculture. *Ecological Economics* 64:253–260.

Zhang, X., R. C. Estoque, H. Xie, Y. Murayama, and M. Ranagalage. 2019. Bibliometric analysis of highly cited articles on ecosystem services. *PLoS One* 14 (2): e0210707. https://doi.org/10.1371/journal.pone.0210707.

Zhu, J. 1999. Local growth coalition: the context and implications of China's gradualist urban land reforms. *International Journal of Urban and Regional Research* 23 (3): 534–548.

Zhu, Wenfeng. 2011. *Study on assessment of agroecosystem service & health and their managerial model in Fujian Province*. Fujian Agriculture and Forestry University.

Zhu, Wenfeng, Songliang Wang, and Claude D. Caldwell. 2010. Agroecosystem service and its managerial essence (in Chinese). *Chinese Journal of Eco-Agriculture* 18 (4): 889–896.

Chapter 4
Agricultural Biodiversity and Food Security

Abstract Biodiversity is essential to ecosystem services but the contributions of agrobiodiversity to agriculture in general and food security in particular are underestimated. Based on the principles of agroecology, this chapter focuses on how to restructure agrobiodiversity, exemplified by technologies applied in the Chinese Ecological Agriculture system dating back to the 1980s.

Keywords Biodiversity · Agricultural biodiversity · Food security · Ecosystem services · The principles of agroecology

Learning Objectives

After reading this chapter, the students should be able to:

1. Discuss the links between biodiversity and ecosystem services.
2. Explain how agrobiodiversity has a key role in food security.
3. Propose a strategy for restructuring agrobiodiversity.

****

> I cannot help thinking that when we finish assigning values to biological diversity, we will find that we do not have very much biological diversity left.
>
> —Erenfeld (1988), cited from Paoletti et al. (1992)

> Bring diversity back to agriculture. That's what made it work in the first place.
>
> —David R. Brower (cited by in Agroecological engineering)

> All flesh is grass.
>
> —Isa. 40:6

It is commonly considered that the world can be categorized into two classes of matter: organic organisms and 'inorganic' surroundings. The former consists of plants, animals, and microbes, which, considered together in a geographic landscape, are termed biodiversity. Biodiversity is defined as (CBD 1992, cited from Mace et al. 2012)

S. Wang, C. D. Caldwell, *Advanced Agroecology*,
https://doi.org/10.1007/978-3-032-08689-1_4

> The variability among living organisms from all sources including, inter alia, terrestrial, marine and other aquatic ecosystems and the ecological complexes of which they are part; this includes diversity within species, between species and of ecosystems.

Simply put, biodiversity consists of the variability of all the species of plants, animals, and microorganisms and the interactions within species, between species and their 'inorganic' surroundings (Toledoa 2006). Because of the interactions among organisms, when ecologists talk about 'surroundings', they use the term 'habitat' on a small scale and environment on a larger scale.

Biodiversity is the complex organizational life format on Earth. It evolves over time and throughout the history of humans. Before agriculture became the main source of human food, our accessors relied on the biodiversity around them for their integrated livelihood. Even after the invention of agriculture, biodiversity in agricultural ecosystems supported the development of the foods upon which we now rely. However, modern industrialized agriculture over the past 100 years has dramatically become a power to threaten biodiversity under increasing pressure from growing populations (Lockwood 1999; Reid 2014). Hence, it is imperative to explore the concept and principles of biodiversity, the relationships among biodiversity, ecosystem services, and human food security.

Discussion started in 2020 regarding global biodiversity targets and there were almost 100 different suggestions in terms of how it might be measured. At present, we need to consider 6 key dimensions: genetic composition, species distribution and abundance, species traits, community composition, ecosystem functioning, and ecosystem structure. If we can measure these six areas, especially over time then we have a good grasp of what is happening to biodiversity.

Research and conservation of biodiversity, which encompasses the variety of life forms from genes to ecosystems, have far-reaching implications. Agricultural biodiversity, a crucial subset, extends these concepts into the realm of food production and farming systems. It is of paramount importance as it underpins global food security by providing a diverse range of crops, livestock, and microorganisms that are resilient to pests, diseases, and climate change. And agricultural biodiversity supports sustainable farming practices by maintaining healthy soil ecosystems, promoting natural pest control, and enhancing nutrient cycling. By exploring and protecting agricultural biodiversity, we can create more sustainable, productive, and resilient agricultural systems for future generations. In this chapter, we extend the significance of biodiversity to agricultural biodiversity and address the losing situation and the pathway for rebuilding the latter both in filed and landscape scale associated with food security.

1 Biodiversity and Ecosystem Services

In 1968, American ecologist Raymond F. Dasmann coined the term 'biological diversity' in his book *A Different Kind of Country* to describe the variety and variability of life forms on Earth (Dasmann 1968). Walter G. Rosen simplified 'biological diversity' as 'biodiversity' in 1985.

Biodiversity refers to the variability among living organisms from all sources, including terrestrial, marine, and other aquatic ecosystems; it consists of three classifications: species diversity, genetic diversity, and ecosystem diversity (Wilson 1988):

- **Species diversity** refers to how many species of organisms live on Earth or in a given area. It is estimated that there are 5–50 million species on our planet, among them 1.4 million named species. Most are insects, plants, and protists/fungi.
- **Genetic Diversity**: The differences in terms of DNA among the individuals of a species. For instance, potato as a species exists in many varieties cultivated in Fujian province of China (Fig. 4.1).
- **Ecosystem Diversity:** Ecosystem diversity refers to the variety of different habitats, i.e. organisms and environments interacting in a given area. Ecosystem diversity also encompasses ecological processes within these habitats with unique physical features, such as climate, soil type, topography, and hydrology. Such habitat varieties become fundamental to genetic and species diversity, without which biodiversity will degrade.

The relationship between biodiversity and ecosystem services is fundamental and multi-layered (Mace et al. 2012). On the one hand, ecosystem services are the results of the interactions between organisms and the environment. Therefore, conserving and managing biodiversity is essential for sustainable provisions of ecosystem services. The loss of biodiversity can lead to a decline in ecosystem services, resulting in significant challenges for human survival and development. On the other hand, ecosystem services provide a material basis and living environment for the maintenance of biodiversity. The various resources and suitable conditions

Fig. 4.1 Genetic diversity of potato species in Fujian province. (Courtesy: Ms. Zhaojuan Zhang, May 22, 2025)

provided by ecosystem services, such as the food, habitat, and water sources in an ecosystem, are crucial for the survival, reproduction, and evolution of numerous species. Once ecosystem services are degraded—such as water pollution in a wetland reducing available clean water—species may face threats of endangerment or even extinction, directly undermining biodiversity.

Recognizing the importance of biodiversity for ecosystem service and human well-being and existence, the United Nations Conference on Environment and Development (the Rio "Earth Summit") issued The Convention on Biological Diversity (CBD) in May of 1992, which came into force on December 29, 1993, after more than 190 countries' signatures and recognitions. The CBD missions are as follows according to the description of the CBD website (https://www.cbd.int/intro):

- The conservation of biological diversity.
- The sustainable use of the components of biological diversity.
- The fair and equitable sharing of the benefits arising out of the utilization of genetic resources.

Economic assessment is the first step to realizing the above-ambitious missions; it is particularly true when linking biodiversity to ecosystem services. Many scientists think that the failure of society to place a value on biodiversity has resulted in the degradation of ecosystem services. They even contribute to the loss of biodiversity itself (Jones-Walters and Mulder 2009). The Economics of Ecosystems and Biodiversity (TEEB) was founded in the European community to provide policymakers with easily digested information since 2008 (TEEB 2008; Brink Ten 2011).

China's Centre for Biodiversity Research valued China's biodiversity in 1998 right after Constanza et al.'s valuing of the global ecosystem in 1997 (Table 4.1).

The data from Table 4.1 offer insight into how to recognize and conserve biodiversity by mainstreaming biodiversity in eco-economic languages:

- The direct use value of China's biodiversity only accounted for a small proportion of the total value, indicating that the current economic market is far from perfect regarding providing an instrument for preserving China's biodiversity.
- The indirect use value of China's biodiversity is far more than the direct use value, showing the large significance of preserving rather than utilizing biodiversity.
- If people must utilize biodiversity, it should be appropriately done with fair compensation of the ecological and social cost of biodiversity utilization.

Under this prevailing surge, the Economics of Ecosystems and Biodiversity (TEEB) was founded in the European community to provide policymakers with easily digested information since 2008 (TEEB 2008; Brink Ten 2011).

Secondly, it is imperative to clarify the connection between biodiversity and human food production (Crist 2017). There are causal effects on two sides: biodiversity as the object of ecosystem goods providing food for human directly on the one hand and food production in the currently prevailed pattern has huge negative effect on biodiversity maintenance on the other hand (Fig. 4.2), implying that

Table 4.1 The monetary values of biodiversity in China in 1998

Value classification	Value types	Economic values ($\times 10^{12}$ Yuan RMB)
Direct use value	Marketing products	1.02
	Other direct use value	0.78
Sum of subclass		1.80
Indirect use value	Organic matter enhancement	23.3
	CO_2 sequestration	3.27
	O_2 release	3.11
	Nutrient cycle and deposit	0.32
	Soil conservation	6.64
	Water preservation	0.27
	Pollutants absorption	0.40
Sum of subclass		37.31
Potential use value	Selecting use value	0.09
	Preserving values	0.13
Sum of subclass		0.22
Total		39.33

Data source: The report on Chinese biodiversity assessment in 1998

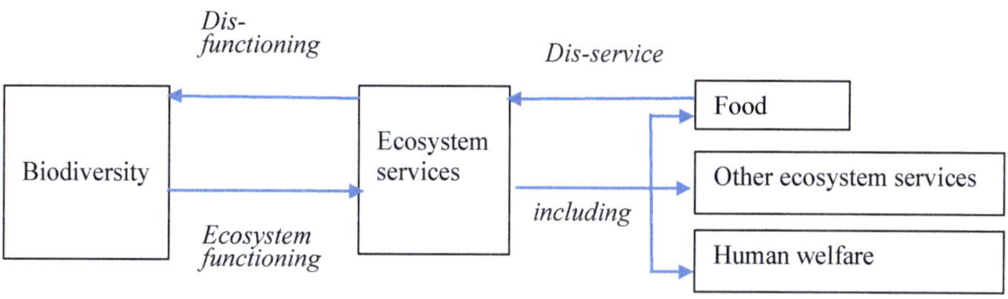

Fig. 4.2 Conceptual model linking biodiversity to ecosystem services connecting with human food and welfare

biodiversity provides the material base for the functioning ecosystem services, the major component of which is human food in terms of human welfare that natural ecosystem provides. However, the industrialized pattern of current food production becomes a disservice and dis-functioning factor for decreasing and degrading biodiversity (Crist 2017).

Numerous studies have shown that food production has had huge impacts on both natural biodiversity and agricultural biodiversity. The expansion of land for crop and animal agriculture contributes to 70% of the deforestation, leading to tremendous biodiversity loss. Agriculture uses 80% of freshwater resources for food production, resulting in more biodiversity existence (Crist 2017). Meanwhile, the overuse of chemicals, including fertilizer and pesticide has directly contaminated the habitats and selectively killed key organisms in the natural ecosystem, degrading the biodiversity in terrestrial and water ecosystems (Sabatiera et al. 2013).

Livestock expansion is increasing globally as the demand for meat grows; this has degraded many species-rich ecosystem especially tropical forest ecosystems where we find the richest biodiversity.

2 Agrobiodiversity and Food Security

Agrobiodiversity is the short form of agricultural biodiversity, meaning the biodiversity in agroecosystems. It is reasonable to conclude that all plants, animals, and microorganisms utilized in agroecosystems are from natural ecosystems even after the artificial modifications, such as domestication, selective breeding, and hybridization (Altieri 1999), but with a much narrower spectrum in scale and size. Among the estimated 30–50 million species on Earth, only 1.3–1.8 million species are scientifically named and described; most are insects, plants, and protists/fungi, the so-called 'decomposers' in function, and most are in the tropics (Paoletti 1992). Among 250,000–750,000 plant species in the world, there are at least 30,000 edible plants; humankind currently only uses approximately 400 plant species for producing human food, medicines, and fibre, for which necessary functions of nutrient cycling, fixation, and fertilization are included. Actually, only 70 plant species within some 12 species of grain crops, 23 vegetable crop species, and about 35 fruit and nut crop species are frequently used for producing food (Fowler and Mooney 1990). 15 plant species produce more than 90% of human food, and three species, wheat, corn, and rice, account for 60% of total human food consumed (Paoletti 1992).

Moreover, like the organisms in natural ecosystems, agrobiodiversity also performs significant ecosystem service beyond the provision of 6'f'—food, fibre, fuel, fodder, feed, and forages, such as nutrient cycling, climate regulation, and pollination.

According to Swift and Anderson (1993), agrobiodiversity could be embodied in the following groups:

- Productive Biota: Crops, trees, and animals chosen by farmers, which play a determining role in the diversity and complexity of the agroecosystem.
- Resource Biota: Organisms that contribute to productivity through pollination, biological control, decomposition, etc., including naturally occurring insects, bacterium, and fungi that can control insect pests and diseases to growing crops and livestock.
- Destructive Biota: Weeds, insect pests, microbial pathogens, etc. which farmers aim to reduce through cultural management and other control methods.

According to Vandermeer and Perfecto (1995), agrobiodiversity can be divided into two types: planned biodiversity and associated biodiversity, the former is the crops and livestock that farmers purposely select and grow in the agroecosystem which is dependent on technologies and management practices (Fig. 4.2); the latter includes all-natural biodiversity associated with crop and livestock habitats such as flora and fauna in farming soil and herbivores, carnivores, decomposers, in a crop field, which provides environmental services to the crops and livestock; however, these are also influenced by human management intensity.

In combination with abovementioned categories, Thrupp Lori Ann (2002) proposed the basic classifications of agrobiodiversity:

- Edible plants and crops which are currently used for producing food and other living materials.
- Livestock and freshwater fish that have been domesticated animal servicing for human meat food.
- Genetic resources consisting of plants and animals in ecosystems that are bred for crop, livestock, and fish.
- Soil organisms which are vital to agricultural production by keeping soil structure and health and providing nutrients.
- Beneficiary organisms including naturally existing insects, bacteria, and fungi that help control pests and diseases affecting crops and livestock.

Unlike biodiversity in a natural ecosystem, the boundary of agrobiodiversity could be broadened, according to Altieri (1999); the factors of agrobiodiversity include:

- The diversity of vegetation and animals within and around the agroecosystem, which are the biological resources existing in natural ecosystems that do good to the varieties in agroecosystems.
- The permanence of the various crops and livestock within the agroecosystem, which refers to crop varieties, and animal hybrids bred for agricultural production.
- The intensity of management in cropping and husbandry system. Those management traditions and strategies usually have certain advantages to sustain production and other ecosystem services; some of them are named as agricultural heritages by international and national organizations.
- The extent of the isolation of the agroecosystem from natural vegetation, referring to those things that only exist in the agroecosystem, is quite different from that of the natural ecosystem.

The last two groups are as important as the first two groups, which can be categorized into agricultural heritages referring to those historical management strategies of agriculture that are precious for agroecological transformation in modern times. For example, intercropping and agri-forestry were two outstanding agricultural heritages from Chinese traditional agriculture with adherence to agroecology principles and closer to concept of sustainable agriculture. These methods were almost discarded in the emerging industrial agriculture, but they can be recalled as models for agroecological transformation (Soldati et al. 2023).

With the expansion of industrial agriculture, simplified cropping systems and intensive inputs of synthesized chemicals have narrowed and degraded agrobiodiversity. On the one hand, the growth of industrialized agriculture erodes the general biodiversity through monoculture and expansion of agriculture to natural habitats, causing the loss of plant germplasm, insects, and soil organisms; on the other hand, extension of a unique crop variety with higher yield leads to crop and livestock genetic loss, according to Thrupp Lori Ann (2000). Since 1900, >90% of the diversity of crops has been lost with the highest of 98% of asparagus and the lowest of 86% of apples. This trend of genetic loss jeopardizes agricultural production and food security.

The loss of agrobiodiversity will result in:

- Crops become vulnerable to pests and disease. Modern monoculture crop systems are increasingly susceptible to pests.
- More frequent disease incidence/transmission. Once infected, such infection is transmitted very quickly and devastates crop production.
- Loss of positive insects, pollinators, beneficial insects, and nutrient-recycling animals decreasing in population, because of habitats damaged by monoculture and chemical input.
- Loss of biodiversity and finally productivity in the soil ecosystem. Soil is an ecosystem with many organisms, their mutual interactions with the micro-environment produce ecosystem services for keeping soil healthy and stabilizing productivity. Long-term intensive chemical input damages the soil structure and habitats which is unfavourable to accommodate biological diversity.

The Irish Potato Famine was an object lesson of the impact of agrobiodiversity loss. The failure of the potato crop in 1846 and 1847, due to potato blight, created a disastrous famine in Ireland. It was, for nearly three and a half million of the eight million population of Ireland, almost their only food as well as their chief means of livelihood. By 1849, almost two million Irish had died or emigrated (Wang et al. 2012).

This lesson demonstrates that agrobiodiversity is fundamental to agricultural production and food security in an area, region, and even in a global scale. According to Paoletti et al. (1992), humans have manipulated 95% of the earth's terrestrial ecosystems (70% earth surface is water cover, and only 30% is terrestrial ecosystems), with 50% of agriculture and pastures for food and fire, 20% of forests for wood, 25% for urbanization and industrialization, to gain 98% of the human food, fibre, and wood. We consume 38.8% of global primary production by the photosynthesis of plants on Earth, and 20% of animal production, which is fed by plant biomass.

The increasing dominance of agriculture that is based on yield and capital has had serious consequences in terms of decreasing agrobiodiversity and food security. Farmer have been forced to give up their 'fork' varieties, those varieties with open seed rights and superior nutritive favours. Not only food quantity was lost but also food quality, economic viability, ecological compatibility, and individual preference were compromised (FAO 1996; Varshney et al. 2021). More than the total food amount being affected, losing biodiversity in agriculture has much to do with food quality, nutritional balance loss due to monoculture, and food diversity loss (Toledoa and Burlingame 2006). In addition, more chemical inputs threaten food safety.

It is time to rebuild agrobiodiversity in agroecosystems.

3 Restructuring Agrobiodiversity Using the Principles of Agroecology

Agriculture has been considered the main destructive force of biodiversity loss (Lockwood 1999; Reid 2014; Scherber 2022), which will have a huge negative impact on human food security, e.g. losing food diversity can lead to diabetes, obesity, malnutrition, and other chronic diseases. However, many scientists consider that conflicts between agriculture and biodiversity can be avoided using agroecology principles and strategies (Reid 2014). Using a suite of agroecological farming techniques can make soils healthier and more productive, as agrochemicals and pollution are diminished, and crop diversity is enhanced, finally making agriculture more resilient and stable.

According to Olivier De Schutter, the UN Special Rapporteur on the Right to Food (de Schutter 2011, cited from Reid 2014):

> Scientific evidence today demonstrates that agroecological methods outperform the use of chemical fertilizers in boosting food production, whilst supporting biological life.

The mechanism is summarized as follows (Knapp and Sciarretta 2023): Agroecology plays a crucial role in protecting, restoring, and promoting agricultural biodiversity through sustainable and resilient farming practices. Traditional intensive agriculture, driven by expanding land use and synthetic inputs, has caused severe biodiversity loss by destroying natural habitats and disrupting ecosystem functions. Agroecology addresses this by adopting "environmentally friendly" methods: diversifying crop production enhances ecosystem services like pollination and nutrient cycling without compromising yields; integrated pest management replaces chemical pesticides with semiochemicals, microbial biocontrol agents, and natural enemy augmentation, reducing non-target species harm and soil degradation.

However, its effectiveness depends on understanding trophic networks across spatial-temporal scales and developing robust indicators to quantify management impacts, as outcomes vary with species ecology and habitat complexity (e.g. pollinators' reliance on surrounding landscapes). Identifying synergies between ecosystem services (e.g. pest control and pollination) and agronomic benefits can increase farmer adoption. Aligned with global targets like the UN Biodiversity Conference (COP 15) and the EU Farm to Fork strategy, agroecology requires scientific evidence to scale practices globally. The new BMC Ecology and Evolution article collection on agroecology aims to synthesize research on sustainable practices, soil management, ecological restoration, and carbon neutrality, fostering healthier agroecosystems and biodiversity conservation.

The following is a historical and a modern example for agrobiodiversity building in China which showed profound agroecological principles in food security practices.

Ancient China's agroecosystem model 'Sangji Yutang' (or dyke-pond system) completely supports Mr. de Schutter's thoughts on biodiversity (Fig. 4.3). This

Fig. 4.3 A built biodiversity agroecosystem called 'Sangji Yutang' landscape in the Pearl River Delta first initiated 1000 years ago in ancient China. (Courtesy: Tang Jianjun, personal bestowing)

initiative dates to the Song Dynasty, which occurred in the Pearl River Delta and the Taihu Lake Basin more than 1000 years ago in China's history. At that time, a magnificent area of wetland with higher water tables made it inappropriate for cropping, our innovative ancestor farmers dug out the soil to create a Pond ('TANG' in Chinese) and its adjacent neighbourhood became a Dyke ('JI' in Chinese). The farmers then cultivated the mulberry on the dykes and grew the fish in the ponds to form an aquaculture-agriculture agroecosystem (AAA), which was also known as a "Mulberry Dyke-Fish Pond" that was developed over centuries, reflecting a profound understanding of ecological principles and resource recycling.

The dyke-pond system is comprised of several interdependent components, showing mutual interaction and interlinking ecological relationships:

- Ponds. It was one of the two main landscapes of the system, in which farmers usually grew different species of fish each eating different feed distributed along vertical line of the ponds, named Chinese poly-aquaculture system by later ecologists. For instance, grass carp near the surface of the pond consume phytoplankton; tilapia, in the middle eat zooplankton. The pond mud was collected occasionally to fertilize the dyke soil. Therefore, while the fish occupied similar spatial niches, they are in separate functional niches.
- Dykes: They surround the pond constructed using the excavated soil from the ponds, which not only prevent water from escaping the ponds but also provide fertile ground for agricultural activities. Farmers cultivated various crops and plants on the dykes, the mulberry trees, vegetables, sugarcane, and other crops. The mulberry trees are worthy of special mention because they were used to feed the silkworms, connecting to the famous Silk Roads. Completing the ecological cycle, the waste products and bodies of silkworms are then used as nutrient-rich feed for the fish.

The dyke-pond system exemplifies a closed-loop agricultural practice, where waste from one component becomes an input for another, thus minimizing waste and enhancing resource efficiency. Key ecological and economic benefits include:

- Nutrient Recycling: Fish waste provides essential nutrients for the crops grown on the dykes, while the vegetation and agricultural by-products serve as feed for the fish. This cyclical nutrient flow reduces the need for external fertilizers and minimizes environmental pollution.
- Biodiversity: The system supports a diverse array of species, both aquatic and terrestrial, contributing to enhanced biodiversity. The varied habitats created by the ponds and dykes offer niches for different organisms, thus promoting ecological balance.
- Economic Resilience: By integrating multiple agricultural and aquacultural activities, the dyke-pond system offers economic resilience to farmers. Diversified production reduces reliance on a single crop or livestock, thereby buffering against market fluctuations and environmental shocks.
- Sustainability: The sustainability of the dyke-pond system is underscored by its low input requirements and high efficiency. The use of local resources and indigenous knowledge ensures that the system remains viable and productive over long periods without degrading the natural resource base.

Another example was conducted by an agroecologist Professor Zhu Youyong in Yunnan province of China in 1990s (Zhu et al. 2000). He used more than 10 varieties of rice intercropping in the same field, the experiment has been conducted for several years and in several township level plots, resulting in resistance to rice blast (*Magnaporthe Grisea Barr*) disease and final decrease in pesticide use of 80% (Fig. 4.4, Zhu et al. 2000). The eco-physiological mechanism is indicated in Fig. 4.5, showing that: (1) a physical barrier hampers bacterium and fungus transmissions; (2) a kind of micro-climate is produced benefiting the rice population growth and development; (3) the hosts for physiological strains of bacterium and fungus are dispersed. All those effects in disease diminish and thus pesticide application decreases.

From a global perspective, the utilization of the principles of agroecology to rebuild biodiversity linked to food security is summarized as follows:

- Build a holistic knowledge network for agrobiodiversity rebuilding sustainable livelihoods including food sovereignty (Toledoa and Burlingame 2006). Because of the complexity of multiple element interaction within and outside agroecosystems, knowledge integration is critical to designing agroecosystems in better structure and function. The advantage of agroecology is its multidimensional thinking and knowledge integration. When facing the objectives of food production and biodiversity conservation, agroecology differs from conventional and capitalizing agriculture models which focus on productivity and economic return and even differs from organic agriculture which focuses almost exclusively on a technology change (De Schutter 2014).
- Reinvent the new (maybe the traditional) farming system which is wildlife-friendly to secure our food supply (Tudge 2014) and could explore a territory integrating agriculture and biodiversity conservation at the same time (Wezel

Fig. 4.4 24 varieties of rice intercropping in Yunnan Province, which decreased pesticide use 80%. (Courtesy: Prof. Luo Shiming, personal bestowing)

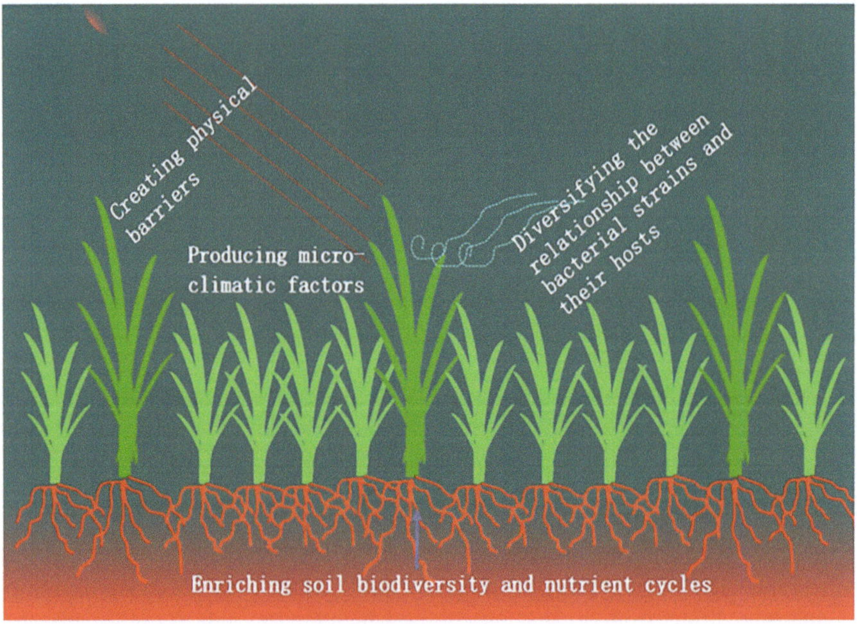

Fig. 4.5 The eco-physical mechanism of rice variety intercropping. (Courtesy: Prof. Luo Shiming, personal bestowing)

et al. 2016; Scherber 2022). A thousand years of history of Chinese Rice-Fish and Rice-Duck Systems have been demonstrated as agroecological territories to embrace production and conservation win–win goals (Figs. 4.3 and 4.4). The eco-economic mechanisms are analysed below: ecologically, in this system, the food chain rice-insect-fish (duck) is set up to decrease fertilizer input, and pesticide input could approach zero, and the fish movement in the field fuels the redox reaction of paddy soil resulting in methane emission decrease; moreover, biodiversity both in the field and under the soil increases and stabilizes the system. Economically, fish (duck) provides diversified production with higher prices that can directly enhance the farmer's income, meanwhile, less fertilizer and almost zero pesticide decreasing the farmer's production costs.

- Offer the key elements in the design of sustainable pest management systems for plant production (Altieri 1993). Because pests harm human economic profit in plant production processes, the appropriate protection measures should be enforced to save the cost without damaging the environment. Therefore, the ecology-based approach to handle pest problems named integrated pest management (IPM) has been a sound and comprehensive strategy. IPM incorporates the simultaneous management and integration of tactics after investigating insect and pest population dynamics to acquire the knowledge of community ecology between pests and natural enemies. Economic thresholds for pest control are used for decisions, with a comprehensive suite of control methods, biological, cultural, mechanical, and chemical to keep agroecosystem's integrity and maintain biodiversity (Pretty and Bharucha 2015; Altieri et al. 2009).
- Rebuild agrobiodiversity for climate change adaptation and mitigation. Agriculture contributes 20–25% of greenhouse gases (GHGs) globally. One includes carbon emissions related to the food consumption phase, this value rises to more than 30% (Ortiz-Bobea et al. 2021; Niu et al. 2024). Table 4.2 lists the sources for carbon emissions in relation to biodiversity loss, indicating that there are two groups and several concrete types of agricultural production activities associated with biodiversity loss that contribute greatly to carbon emissions. The GHGs of the emission include carbon dioxide, methane, and nitrous oxide. This implies that decreasing such activities could diminish biodiversity loss and at the same time mitigate GHGs emissions, or rebuilding biodiversity in agricultural filed on one's own initiative, for example, conservation tillage and more rice varieties intercropping in paddy field will definitely make agriculture more adapted to climate change (Figs. 4.6 and 4.7).

Table 4.2 The main sources for carbon emissions in relation to biodiversity loss

Group of agricultural activities	Type of agricultural activities	Main GHGs emissions
Greenhouse gas emissions from agricultural production	• Burning crop residues • Crop residues • Rice cultivation • Savanna fires • …	• CH_4, N_2O, CO_2 • N_2O • CH_4, N_2O • CH_4, N_2O • …
Land-use change-related emissions	• Net forest conversion • Fires in humid tropical forests • Fires in organic soils • …	• CO_2 • CO_2 • CH_4, CO_2 • …

Fig. 4.6 A thousand years of history of Chinese rice-fish systems. (Courtesy: Tang Jianjun, personal bestowing)

Fig. 4.7 A thousand years of history of Chinese rice-duck systems. (Courtesy: Tang Jianjun, personal bestowing)

- Calculate the intrinsic value of agrobiodiversity to mainstream its protection process in economic accounting (Lanka et al. 2017; Voglhuber-Slavinsky et al. 2023). The current industrial agriculture has compromised traditional agroecosystems and replaced biological functions, originally provided by diverse communities of organisms. Agroecology calls for the holistic management of agroecosystems. But in the current economy-driven society, everything is measured by monetary value; if agroecological function cannot be monetized, the function and practice will be in vain. Agroecology has developed an accounting system for agrobiodiversity that paves the way to rebuilding agricultural biodiversity for sustainable livelihoods (also see Chap. 3 of this book in detail on how to calculate and monetize the ecosystem services for agri-food system).

Another outstanding case from urban area of Mexico that is summarized in Box 4.1.

Box 4.1 Urban agroecology enhances agrobiodiversity and resilient, biocultural food systems. The case of the semi-dryland and medium-sized Querétaro City, Mexico (Villavicencio-Valdez et al. 2023)

- **Background**

  Global challenges such as agricultural biodiversity loss, diet-related health crises, and urbanization-driven ecological degradation highlight the need for resilient food systems. Industrial agriculture has eroded 70% of crop genetic diversity, while processed food dominance exacerbates obesity and diabetes, with 70% of Mexico's population facing food insecurity. Medium-sized cities like Querétaro, Mexico, face compounded issues: rapid sprawl encroaches on agricultural lands, groundwater depletion reaches −105.9 Mm^3 annually, and 40% of its residents' experience food insecurity. Urban agroecology (using agroecological principles for urban agriculture), though unrecognized in official statistics, offers potential to enhance food sovereignty and socioecological resilience.
- **Processes**

  Diversified crop planting: Cultivating a wide range of plant species (e.g. chilli, aloe vera, chard, and prickly pear) to enhance agrobiodiversity, with some gardens hosting up to 86 varieties. Polycultures and perennial-dominated systems, such as integrating drought-tolerant CAM succulents (Opuntia, aloe), are emphasized for climate resilience.

  Sustainable soil and water management: Composting and biomass recycling to improve soil fertility and organic matter. Rainwater harvesting (e.g. Tlaloque systems) and grey water reuse from kitchens for irrigation. Mulching, contour planting (in peri-urban areas), and vegetation integration to retain soil moisture and prevent nutrient loss.

  Ecological Pest control: Using natural predators, intercropping, and flower borders to attract pollinators and beneficial insects, reducing reliance on synthetic pesticides.

(continued)

Box 4.1 (continued)

Social and cultural practices—Biocultural memory preservation: Incorporating traditional knowledge from rural migrants, such as native crop domestication, seed exchange, and ancestral cooking practices. Community-led initiatives: Establishing community gardens, school gardens, and 'urbicultura' networks for food production, social inclusion, and knowledge sharing (e.g. 'De urbicultor a urbicultor' dialogues).
Adaptive garden design—Optimal plot size: Gardens of approximately 200 m^2 show the highest agrobiodiversity and productivity, balancing manageable scale with ecological complexity. Vertical gardening, raised beds, and reused containers: Innovating space use in urban environments, including rooftops, vacant lots, and public parks.
Reduced dependence on external inputs: Relying on organic fertilizers, local seed varieties, and decentralized production to minimize fossil fuel use and enhance self-sufficiency.

- **Achievement**
These practices collectively promote socioecological resilience, food sovereignty, and climate adaptation in urban and peri-urban settings. The agroecological management significantly enhances species richness ($P < 0.001$), with urban gardens averaging 86 plant varieties—higher than peri-urban sites due to non-commercial, culture-driven planting. Key species like *Capsicum annuum* (chilli), *Aloe vera*, and *Beta vulgaris* (chard) dominated, while drought-tolerant CAM succulents (*Opuntia*, *aloe*) showed high IVI, underscoring climate resilience. Gardens of ~200 m^2 demonstrated optimal biodiversity and productivity, balancing manageability and ecological function. Socioecologically, UA fosters resilience through nutrient recycling, water conservation, and community cohesion. Women-led gardens (64% of participants) preserve biocultural memory, such as traditional seed exchanges and collective cooking, though marginalized communities face displacement risks (e.g. the 'El Huerto del Buen comer' garden lost to urban development). Policy implications include advocating for secure land tenure, integrating 200 m^2 garden models into urban planning, and supporting community-driven knowledge sharing (e.g. 'De urbicultor a urbicultor' dialogues). The research highlights UA as a scalable, grassroots solution to enhance food security, biodiversity, and social inclusion in Latin American cities, emphasizing the need for adaptive governance and inclusive policies.

In summary, agroecology, agricultural biodiversity, and food security (amount, nutrition, safety and distribution, and preferences) are interdisciplinary issues that need to be developed into innovative agroecology networking systems with novel ecological theories (Breckling and Reuter 2004), such as an ethno-agroecology (a citizen agroecology, Altieri 1993), a farmer-led agroecology (Bohan et al. 2022), and participatory action (agroecology) research (Nimmo et al. 2023). Such integration of agroecology education towards the nexus of biodiversity and food security is needed.

References

Altieri, Miguel A. 1993. Ethnoscience and biodiversity: key elements in the design of sustainable pest management systems for small farmers in developing countries. *Agriculture, Ecosystems and Environment* 46 (1–4): 257–272.

Altieri, Miguel A. 1999. The ecological role of biodiversity in agroecosystems. *Agriculture, Ecosystems and Environment* 74:19–31. 2009.

Altieri, Miguel A., Clara I. Nicholls, and Luigi Ponti. 2009. Crop diversification strategies for pest regulation in IPM systems. In *Integrated pest management: concepts, tactics, strategies and case studies*, ed. E. B. Radcliffe, W. D. Hutchinson, and R. E. Cancelado, 116–130. Cambridge, UK: Cambridge University Press. ISBN 978-0-521-87595-0.

Bohan, David A., Anett Richter, Miranda Bane, Olivier Therond, and Michael J. O. Pocock. 2022. Farmer-led agroecology for biodiversity with climate change. *Trends in Ecology & Evolution* 37 (11): 927–930.

Breckling, Broder, and Hauke Reuter. 2004. Analyzing biodiversity: the necessity of interdisciplinary trends in the development of ecological theory. *Poiesis & Praxis* 3:83–105.

Brink Ten P. 2011. The economics of ecosystems and biodiversity in national and international policy-making. Washington, DC: Earthscan

Crist, Eileen, Camilo Mora, and Robert Engelman. 2017. The interaction of human population, food production, and biodiversity protection. *Science* 356:260–264.

Dasmann, Raymond F. 1968. *A different kind of country*. 1st ed. London: Collier-Macmillan.

de Schutter, Olivier. 2014. *Agroecology and the right to food*. Interview by Margriet Goris. http://leisaindia.org/wp-content/uploads/2014/07/LEISA-India-June-2014-Page-18-19.pdf. Accessed March 24, 2025.

FAO. 1996. *Rome Declaration on World Food Security and World Food Summit Plan of Action*. https://www.fao.org/3/W3613E/W3613E00.htm. Accessed June 13, 2024 in Vancouver.

Fowler, C., and P. Mooney. 1990. *Shattering: food, politics and the loss of Genetic Diversity*, 278. Tucson, AZ: University of Arizona Press.

Jones-Walters, L., and I. Mulder. 2009. Valuing nature: the economics of biodiversity. *Journal for Nature Conservation* 17:245–247.

Knapp, Jessica, and Andrea Sciarretta. 2023. Agroecology: protecting, restoring, and promoting biodiversity. *BMC Ecology and Evolution* 23:29. https://doi.org/10.1186/s12862-023-02140-y.

Lanka, Sanjay V., Iqbal Khadaroo, and Steffen Böhm. 2017. Agroecology accounting: biodiversity and sustainable livelihoods from the margins. *Accounting, Auditing & Accountability Journal* 30 (7): 1592–1613.

Lockwood, Jeffrey A. 1999. Agriculture and biodiversity: finding our place in this world. *Agriculture and Human Values* 16:365–379.

Mace, G. M., K. Norris, and A. H. Fitter. 2012. Biodiversity and ecosystem services: a multilayered relationship. *Trends in Ecology & Evolution* 27:19–26.

Nimmo, Evelyn Roberta, Erin Nelson, Laura Gómez-Tovar, Mariol Morejón García, Andrew Spring, André E. B. Lacerda, Alessandra Izabel de Carvalho, and Alison Blay-Palmer. 2023. Building an agroecology knowledge network for agrobiodiversity conservation. *Conservation* 3:491–508.

Niu, Kun Yu, Ruo Hao Ge, Mei An Chen, Shuqin Jin, and Jing Liu. 2024. Structural characteristics, development trends, and international comparison of greenhouse gas emissions in China's agri-foodsystem under the dual-carbon objectives (in Chinese). *Scientia Agricultura Sinica* 57 (21): 4290–4307.

Ortiz-Bobea, Ariel, Toby R. Ault, Carlos M. Carrillo, Robert G. Chambers, and David B. Lobell. 2021. Anthropogenic climate change has slowed global agricultural productivity growth. *Nature Climate Change* 11:306–312. https://doi.org/10.1038/s41558-021-01000-1.

Paoletti, M. G., D. Pimentel, B. R. Stinner, and D. Stinner. 1992. Agroecosystem biodiversity—matching production and conservation biology. *Agriculture, Ecosystems and Environment* 40:3–23.

Pretty, Jules, and Zareen Pervez Bharucha. 2015. Integrated pest management for sustainable intensification of agriculture in Asia and Africa. *Insects* 6:152–182. https://doi.org/10.3390/insects6010152.

Reid, Vanessa. 2014. Agriculture, agroecology, and biodiversity. *Biodiversity* 4:239–240.

Sabatier, Rodolphe, Katrin Meyer, Kerstin Wiegand, and Yann Clough. 2013. Non-linear effects of pesticide application on biodiversity-driven ecosystem services and disservices in a cacao agroecosystem: a modeling study. *Basic and Applied Ecology* 14:115–125.

Scherber, Christoph. 2022. Reconciling biodiversity and production in farming systems. *Basic and Applied Ecology* 65:62–66.

Soldati, Cristian, Anna Irene De Luca, Nathalie Iofrida, Emanuele Spada, Giovanni Gulisano, and Giacomo Falcone. 2023. Ecosystem services and biodiversity appraisals by means of life cycle tools: state-of-art in agri-food and forestry field. *Agriculture & Food Security* 12:33.

Swift, M. J., and J. M. Anderson. 1993. Biodiversity and ecosystem function in agroecosystems. In *Biodiversity and ecosystem function*, ed. E. Schultze and H. A. Mooney, 57–83. New York: Springer.

TEEB. 2008. *The economics of ecosystems and biodiversity—an interim report*. Brussels: European Commission.

Thrupp, Lori Ann. 2000. Linking agricultural biodiversity and food security: the valuable role of agrobiodiversity for sustainable agriculture. *International Affairs* 72:265–281.

Thrupp, Lori Ann. 2002. Agricultural biodiversity: a key element of ecosystem health and sustainable food security. In *Managing for healthy ecosystems*. CRC Press.

Toledoa, Alvaro, and Barbara Burlingame. 2006. Biodiversity and nutrition: a common path toward global food security and sustainable development. *Journal of Food Composition and Analysis* 19:477–483.

Tudge, Colin. 2014. Agroecology and biodiversity: the kind of farming we need to secure our food supply is also wildlife friendly. *Biodiversity* 15 (4): 290–293. https://doi.org/10.1080/14888386.2014.977349.

Varshney, Rajeev K., Abhishek Bohra, Manish Roorkiwal, Rutwik Barmukh, Wallace A. Cowling, Annapurna Chitikineni, Hon-Ming Lam, Lee T. Hickey, Janine S. Croser, Philipp E. Bayer, David Edwards, José Crossa, Wolfram Weckwerth, Harvey Millar, Arvind Kumar, Michael W. Bevan, and Kadambot H. M. Siddique. 2021. Fast-forward breeding for a food-secure world. *Trends in Genetics* 37 (12): 1124–1136. https://doi.org/10.1016/j.tig.2021.08.002.

Vandermeer, J., and I. Perfecto. 1995. *Breakfast of biodiversity: the truth about rainforest destruction*, 185. Oakland: Food First Books.

Villavicencio-Valdez, Gabriela Valeria, J. Jacobi, M. Schneider, M. A. Altieri, and H. Suzán-Azpiri. 2023. Urban agroecology enhances agrobiodiversity and resilient, biocultural food systems. The case of the semi-dryland and medium-sized Querétaro City, Mexico. *Frontiers in Sustainable Food Systems* 7:1066428. https://doi.org/10.3389/fsufs.2023.1066428.

Voglhuber-Slavinsky, Ariane, Nahleen Lemke, Joseph MacPherson, Ewa Dönitz, Mathias Olbrisch, Philipp Schöbel, Björn Moller, Enno Bahrs, and Katharina Helming. 2023. Valorization for biodiversity and ecosystem services in the agri-food value chain. *Environmental Management* 72:1163–1188. https://doi.org/10.1007/s00267-023-01860-7.

Wang, Songliang, C. D. Caldwell, S. L. Kilyanek, and J. Sziklay, eds. 2012. *Agroecology*. Beijing: China's Science Press.

Wezel, A., H. Brives, M. Casagrande, C. Clément, A. Dufour, and P. Vandenbroucke. 2016. Agroecology territories: places for sustainable agricultural and food systems and biodiversity conservation. *Agroecology and Sustainable Food Systems* 40 (2): 132–144.

Wilson, E. O. 1988. The current state of biological diversity. In *Biodiversity*, ed. E. O. Wilson, 3–18. Washington, DC: National Academic Press.

Zhu, Y., H. Fen, Y. Wang, Y. Li, J. Chen, L. Hu, and C. C. Mundt. 2000. Genetic diversity and disease control in rise. *Nature* 406:718–772.

Chapter 5
Agroecology and Agri-Food System Transformation

Abstract The predominant conventional agriculture in the world responds to the production (supply) and distribution (demand) cycle, but perhaps pays too much attention to the production and neglects distribution. Agroecology can bridge the gap between production and distribution, as well as helping define policy and technologies for a sustainable global food system. This chapter addresses the challenges of transforming the global agri-food system and agroecology's role in such urgent transformation.

Keywords Agroecology · Agroecological transformation · Global food system · Agri-food system transformation · Value chain

Learning Objectives
After reading this chapter, students should be able to:

1. Critically evaluate the challenges facing the global food systems.
2. Explain why there is a need to integrate the parts of agriculture and food systems into a whole agri-food system.
3. Discuss the role of agroecology in global agri-food system transformation.
4. Describe a possible framework for using agroecology to rebuild a sustainable food system.

****

> The problem with agriculture is a problem with culture.
> —Wendal Berry (1977)

> Supermarket prices don't reflect the real cost of supplying food for the family table.
> —Dana L. Jackson and Laura L. Jackson (2002)

Without a doubt, food is the most important material service for human survival and development, and every country places food as the top priority among other affairs of state. However, throughout human history, there has been a struggle to match food production and distribution. In this sense, there is an opportunity to improve

S. Wang, C. D. Caldwell, *Advanced Agroecology*,
https://doi.org/10.1007/978-3-032-08689-1_5

the current global food system, with better integration of production and distribution sides and with better governance.

Almost at the same time as the world-scale financial crisis developed in 2008, there was an 'unprecedented crisis' in the global food system. This is evidenced by food prices and a historical peak in world hunger since the twentieth century. Since then, many experts and politicians have proposed various solutions to the issues, but technological or superficial interventions have provided less than optimum results (Kirschenmann 2007). In 2011, Schutter Olivier provided a special report to the UN human rights committee called 'Agroecology and the Right to Food'. The report positioned food as a human right and stated that utilizing the principles of agroecology as the tool can solve the problem (de Schutter 2011a, b, 2012). This chapter will combine disciplinary, scientific, and technological aspects of agroecology to address such systematic millennium food problems.

1 Global Food System: Production and Distribution Gaps and Conflicts

For the near future, world population will continue to grow. Even though the birth rate has started to decline both in developed and developing countries, analysts estimate that the world population will reach 9 billion in 2050 (Wezel and David 2012; World Population Prospects 2024, https://population.un.org/wpp/) and 11.2 billion at the end of this century (Crist et al. 2017; World Population Prospects 2024, https://population.un.org/wpp/). With a 12.5% increase from the current number of 8 billion, food production needs to increase 70% by the end of 2050 and double or even triple by the end of 2100 (Crist et al. 2017). This will pose a great deal of pressure on our agriculture and food system (hereafter referred to as agri-food system) ecologically, economically, and sociologically:

- **Pressures on the Future Environment:** For most of the 10,000 years of farming history, the production and consumption of food have been intimately connected and coordinated. This harmonious relationship was broken nearly 200 years ago with the onset of marketing and capitalization of the agri-food system. An industrial paradigm was utilized to maximize yields with increases in external inputs; this has resulted in the loss of soil and biodiversity, pollution of soil, water, and food itself, and detrimental effects on human health (Pretty 2003). To feed the nine billion people in 2050, it is estimated that 10^9 hm^2 of natural land will be converted into agricultural land, resulting in a 2.4–2.7-fold increase in nitrogen and phosphorus-based eutrophication to the ecosystem (Tilman et al. 2001), which further imposes more pressure on global environment and resources. With the present production model, pesticide use must increase; this will lead to increases in ecosystem destruction and food safety problems. Therefore, facing the challenges of food production in 2050, a degraded environment will make it difficult to meet the demands (Springmann et al. 2018). In addition, global warming will cause difficulties for the whole food system, including production, dis-

tribution processing, and consumption. At present the agri-food system contributes 19–29% of the atmosphere GHGs emissions and food production accounts for 80–86% of that (Vermeulen et al. 2012). Food distribution and consumption account for the rest of 14–20%. Consequently, climate change in turn affects food production in various ways, for instance, quicker disease transmission further damages environment and food safety, damaging the goal of food production processes themselves (Dale 2020).
- **Pressures on the Future Economy:** The world agri-food system is under great pressures economically; this is reflected in (1) countries which are dependent on an export-oriented agricultural economy which does not fit the needs of the people. It results in a gap in production and needs in an unrealistic 'Free trade' system. This often results in international movement of 'food aid' to fill the gaps (Rosset 2011; William 2012; Ramankutty and Dowlatabadi 2021). Such a system, in reality, does not reach its original goal; rather, it exacerbates hunger, when capitalized companies exploit land in poor countries to produce and export food back to the home countries. Poor consumers in these countries rely on cheap imported food from another poor country and the national food system in both countries is destroyed, along with the health of the consumers. (2) Several large capitalized companies control the agri-food system and their goal is to provide more profit for their stakeholders, on the backs of consumers with higher prices. In the meantime, small-scale farmers lose their competitiveness due to unaffordable cost of the means of production, and they share less and less in the value chain (IPES-Food 2017). Therefore, there is a need for massive redesign of the economic structures of the global agri-food systems rather than just partial modification.
- Pressures on Future Society: There is a huge gap between production and distribution of food under the current unsustainable agri-food system (Gliessman 2011a); unfortunately, food aid by several non-government organizations is not the final answer. The United Nations established the World Food Programme (WFP) in charge of food aid in 1976. After 50 years of operation, we have finally realized that the WFP meant to serve poor consumers is like the 'fish' versus 'fishing' analysis. It is far better to help local farmers to feed their populations over the long term rather than using aid to encourage them to continue on the treadmill of the exporting and importing model (Rosset 2011). In this sense, all old routes need to be changed entirely under the flagship of food sovereignty, a person's natural right, every person born with accessible, adequate, affordable, and healthy food (Martínez-Torres and Rosset 2010; Dale 2023).

2 Challenges for Agri-Food Value Chain Transformation

Because of the huge successes in the past of capitalized and industrialized agri-food systems, people, including politicians, scientists, managers, and CEOs (chief executive officers) of big agricultural companies tend to be addicted to such a fragmented and compartmentalized agri-food model. This hampers the agroecological transformation of the agri-food system and its economic restructuring (IPES-Food 2017; Biovision Foundation for Ecological Development and IPES-Food 2020). For instance, the Green Revolution period starting from the 1960s, which was so successful in breeding and extending the semi-dwarf and productive grain crop varieties, has been seen as an excellent model for the future. When facing ever more serious food insecurity problems, people think the answer is a second Green Revolution based on the old model. They are blinded to its hugely negative effect on environment degradation, human health, and agri-food sustainability; even adding new biotechnology and informatics technology will not be effective. The power and influence of large corporations in agri-food supply chains have had a significant effect on agricultural production and land use. This has resulted in some initiatives away from human food-related activities and towards biofuels, seeking to make trade-offs between food security and energy needs. In the meantime, public sector investment in agri-food system has declined, moving the world agri-food system into a capitalized/industrialized feedback loop (IPES-Food 2016). The feedback factors are:

- **Concentration of Power:** Traditional agri-food systems rely on small-scale farmers to produce and supply most national agricultural products and food, even in the current industrialized agri-food system. Small-scale farmers still provide 60% of the world food supply (IPES-Food 2017); however, such contribution of the small-scale farmers exists under the growing trend of the concentrated power of capital. Consequently, small-scale farmers either tend to be bankrupt or quit their land for city labour; this, in turn, consolidates the concentration of power of capital by concentrating the arable landholdings, monopolizing the market (including product market and means of production market), and hampering the agroecological transformation of the agri-food system. Worldwide, the four companies, Archer Daniels Midland, Bunge, Cargill, and Louis Dreyfus, occupy 60% of the grain food market, and in China, large state-owned companies and companies with hybrid ownership such as Longping High Tech, Beidahuang, and Shandong Denghai maintain a controlling presence in China's grain seed industry. They tend to maintain the status quo of the agri-food system. Therefore, it is understandable that the state's agriculture subsidy programs are in favour of large-scale agri-food companies, and institutes and universities' agriculture research is also biased towards industrial agriculture. A feedback loop is formed.
- **'Feed the World' Narratives:** There will be more mouths to feed with the population increase even with a very low birth rate in the world; with this in mind, most countries misinterpret food security merely as the amount of grain or staple food, based on which they make food security policy in favour of industrialized agriculture. However, food security does not merely refer to food amount let alone grain production, it embraces every aspect of the human diet, including quantity, quality, and safety, even individual preference (FAO 1996; HLPE 2020). Food is more than production, distribution is also an elephant in the room

of food security and sometimes distribution is the true key to security. Therefore, the industrial system of production does not solve the problem, on the contrary, it has added the problems of food safety and inequity in the value chain. In 2020, 811 million people went to bed without food and 2 billion people suffered from micronutrient deficiency (FAO 2021). In China, there is significant misunderstanding of food security. The 'Feed-China' narrative has been a top priority policy and practice since the question of 'Who will feed China' proposed by Lester Brown in 1994 (Brown 1995). Such a productivity-centred narrative justifies the continuation of the industrial agricultural system, deflects attention away from its consequences, and side-lines the discussion of the latent benefits of diversified agroecology systems. As a result, the protection of 1.2×10^8 hm^2 farmland has been a slogan conflicting with quicker industrialization and urbanization, in practice using marginal land to replace the good quality land in urban and suburban areas has become an inevitable option. In the meantime, the expansion of industrialized agriculture further degrades and contaminates the arable land by using four times fertilizer and two times more pesticide than the world's average. Both become liabilities to the feed-China narratives, ironically, feed-China narratives create this consequence.

- **The Expectation of Cheap Food:** Using mass advertising and consumer manipulation, food retailers have produced a generation of consumers who are open to the need for a vast variety of cheap foods and uniform commodities. This is complemented by the production of a wide variety of processed, relatively cheap mass-produced fast food. This is reconfigured consumption patterns worldwide and encourages the increasing demand of animal protein which supports the industrial livestock reduction system. Therefore, if there is to be a food system transformation based on agroecological principles, consumers' cognitive competence and empowerment is the first thing to be addressed via ecology education.
- **Compartmentalized and Short-Term Thinking:** Worldwide, because of the prevailing social division of labour, the agri-food system has not become an integrative industry compared to other industries in our society. Production happens in rural areas, processing in suburban areas, and trading in cities. Correspondingly, there exists compartmentalization in agricultural research and education in agriculture-associated universities and institutes. This compartmentalized thinking and industrialized agri-food approach interact as both cause and effect, producing again a feedback loop. In such a system, firstly, food productivity is isolated from sustainability of production, food accessibility, and food utilization; secondly, food production is isolated from food distribution. Finally, people think only in the short-term, prioritizing profit as the unique goal of each siloed link of agri-food system. Local government needs better short term performance, and investors are interested in acquiring quick investment returns; this results in government support for large farms for quick improvements and marginalized diversified agroecology approaches of farming since they do not generate immediate benefits.

3 Agroecology and Agri-Food System Transformation

After addressing the imperative challenge of the value chain of global agri-food system, agroecology which is grounded in the marriage of ecology with agronomy and food system sciences has emerged as a holistic path to find a way out (Gliessman 2011a, 2020; de Schutter 2011a, b), but how is such a link established?

Connecting to the evolution of the concept of agroecology explained in Chap. 1, after entering the new millennium, a new cornerstone of agroecology definition appeared in 2003, when Francis et al. concluded that the current fossil fuel-oriented agriculture and food system is in no way sustainable, due to the compartmentalized disciplines and sciences that are instructing its development. They expanded the concept of agricultural system to food system and speculated ecology as a holistic science and discipline that should be integrated into the latter, such integration made agroecology the ecology of the entire food system (Francis et al. 2003). In response to Francis et al. (2003) 's assertion, Stephen R. Gliessman reedited his agroecology textbook with the subtitle 'the ecology of sustainable food systems' (Gliessman 2003) after co-authoring Francis's leader paper (Gliessman 2011a).

The subsequent cornerstone of agroecology definition appeared in 2009, Alexander Wezel and his colleagues proposed reframing agroecology as a science, a movement, and a practice (Wezel et al. 2009). Such definition is not only driven by the agroecology practices mainly in North America (Wezel and David 2012) but also recognized by the United Nations. In 2011, Oliver de Schutter, the special rapporteur on the Right to Food of the UN submitted the report named 'Agroecology and Right to Food' to the UN Office of the High Commissioner for Human Rights and was approved. The report stated that agroecology could facilitate the global sustainable food system in five ways (de Schutter 2011a, b):

- Agroecology increases food availability by raising the productivity at the field level.
- Agroecology enhances food accessibility by reducing poverty from socioeconomic perspective.
- Agroecology elevates food adequacy by contributing to food nutrition improvement by enhancing agrobiodiversity and agrodiversity (see Chap. 4).
- Agroecology assists food system sustainability by contributing to climate change adaptation of agri-food system.
- Agroecology empowers farmers by involving participation in practices (see detail in Chap. 8).

When interviewed about issues of sustainable food systems at a conference in Wageningen University and Research (WUR), the Netherlands in 2014, Oliver De Schutter concluded the role of agroecology in rebuilding a sustainable food system globally and locally (de Schutter 2014):

> Agroecology is really common sense. It means understanding how nature works, to replicate the natural workings of nature on farms in order to reduce dependency on external inputs. Agroecology preserves the ability for future generations to feed themselves. I

believe we should teach more about agroecology and encourage exchanges between farmers. We cannot continue in this impasse of an oil dependent food production system.

All in all, he concluded that agroecology is a tool for the realization of the right to food (de Schutter 2012) with three scale levels as a science, as a practice, and as a social movement (Wezel et al. 2009). The narrative and cases are listed as follows:

- Agroecology, as a science, refers to the science and technology aspect of agriculture and food system transformation. Agroecology first is the science of developing new food production technologies that could move agri-food systems towards a real transition; thus, such transition is called agroecological transformation. Linguistically, agroecology means the marriage of the sciences of agronomy and ecology, resulting in a strong grounding in the science of ecology for both the environment and people. But don't be overly reliant on this technological aspect and scientific rhetoric of agroecology, whatever you think the sound sciences and technologies are, you would miss something without the holistic image of the transformative initiative for the current agri-food system (Gliessman 2011a, 2019a, b). Vandermeer et al. (2018) addressed the holistic connotations of agroecology or ecology of the entire food system (also in Gliessman 2019a, b):
  - **The Ecology of Agroecosystems:** This is the basic connotation of agroecology defined as a science by Wezel et al. (2009), because scientifically agroecology addresses the principles of ecology that are embedded in the relationship between agricultural organisms and their environment, such dynamics of relationship are termed agroecosystems, embracing all parts of agriculture. Essentially, the natural science parts of agroecology are to manage agroecosystem properly by means of ecological thinking and principles (Vandermeer 2011).
  - **Equity in Global and Local Food Systems:** When considering agriculture is more than food production, distribution and consumption are also essential parts of agriculture, then agroecosystems could be extended to the agri-food system. In such a broader system, equitably shared resources and outcomes for people or stakeholders are mandated. Most of the current crises of the agri-food system are driven from unequal distribution of resources and outcomes of agri-food systems. Therefore, agroecology, both defined as a practice and a social movement, should be expanded to social sciences aiming at building a sound eco-social system (Rahman et al. 2021).
  - **Cultural Dimensions of Agri-Food System**: Culture is the heart of food both from the historical development of agriculture and from the psychological essence of agriculture. As Jules Pretty (2002) stated in his thought-provoking work named agriculture without culture, considering food as solely commercial production in the modern industrial and capital agricultural system leads to a dangerous crossroad (Morgan and Trubek 2020). The currently urgent transformation of our agri-food system towards agroecology should take culture back inside agriculture (Vandermeer et al. 2018; Gliessman 2020), in doing so, one way is to take agriculture and food as public good instead of

economic commodity (Jackson 2002). Another way is to highlight cultural services of agri-food systems to embody higher monetary value returning to farmers and the rural landscape (Jackson 2002).

- **Human Health**: Food is also the basis of human health (Deaconu et al. 2021); the ultimate goal of the agri-food system is to provide healthy, safe, nutritious food with diverse nutrient contents good for the human body preventing malnutrition.

As Gliessman (2019a, b) commented:

> If researchers were to embrace this framework as a research paradigm, their work would undoubtedly greatly accelerate the much-needed transitions toward global food system sustainability.

In this sense, Gliessman (2019a, b) continued to state:

> Agroecology is an interdisciplinary space where folks from different disciplines can contribute their disciplinary perspectives to a common goal. But agroecology is also transdisciplinary, not only as a place for the normal disciplines of science to come together, but it is a place where the deep intergenerational knowledge of farmers and consumers joins with these disciplines and new knowledge about food system sustainability comes about.

By doing this, the culture of agriculture can be reshaped (Ferguson et al. 2019) and multi-functions of agriculture can be brought back (IPES-Food 2015).

Fortunately, under the influence of FAO in agroecology strategic planning since 2011, a great deal of agroecology practice and agri-food transformative cases has emerged that could be extended worldwide, IPES-Food (2018). By using agroecology technology, the agri-food system in local landscape is transformed to a great degree. Among those seven cases, one on turning strawberry monoculture farming into multi-cropping farming of Jim Cochran of Swanton Berry Farm in Santa Cruz, California, USA, can help us understand why agroecology is a science of developing new technologies for agri-food transformation. Growing strawberries was the main cropping project in Jim Cochran of Swanton Berry Farm for a long time via intensive agrochemicals input under the conventional agriculture mode before 1980. Under the advice and instruction from the Agroecology Program in the University of California at Santa Cruz (UCSC), as well as in response to regional market demand, the farm decided to transition into farming strawberries organically in the early 1980s. It was very hard for the transition, which needed a series of organic management practices in place of accustomed agrochemical and industrial modes at the beginning. Among those new practices, crop rotation substituted for monoculture; a 'push-pull' pest management technique, plus with releasing predator mites onto an organic strawberry field substituted intensive pesticide spraying. A 'push-pull system' has been used for pest and weed management in Kenya, consequently, maize yields and milk production have doubled. This system uses 'pushing' away pests from corn by interplanting *Desmodium*, which is used as fodder for dairy cows, at the same time, uses *Napier* grass, which secretes a sticky gum that traps insects for 'pulling' pest (Khan et al. 2011, see Box 5.1).

Box 5.1 'Push-Pull' Pest Management Technique (IPES-Food 2016, 2018; Muramoto et al. 2014; Swezey et al. 2013; Khan et al. 2011)

Push-pull systems represent sophisticated and integrated approaches to managing pests, weeds, and soil health within agricultural ecosystems. In the context of mixed cereal-livestock farming systems, these systems demonstrate a strategic combination of attraction and repulsion mechanisms. Specifically, stem-borers are drawn towards *Napier* grass, which serves as the "pull" component. Simultaneously, a repellent legume, *Desmodium*, intercropped with the primary cereal crop acts as the "push" element, effectively deterring stem-borers from the cereal plants.

The utility of *Desmodium* extends beyond pest repulsion. Its root exudates play a crucial role in controlling the parasitic striga weed by inducing abortive germination, thereby suppressing the weed population. Moreover, *Desmodium* contributes significantly to soil fertility enhancement. Through the process of nitrogen fixation, it enriches the soil with this essential nutrient. Additionally, it provides natural mulching, increases biomass, and helps mitigate soil erosion. Both Napier grass and *Desmodium* offer the added benefit of serving as high-quality animal fodder, which not only supports milk production but also diversifies the income streams of farmers.

Mustard cover crops have been the subject of scientific investigation due to their potential to reduce weed infestations and disease prevalence through allelopathic mechanisms. These crops release toxic natural compounds into the environment, which inhibit the growth and development of competing weeds and pathogens. Similarly, broccoli has emerged as a pivotal rotation crop. It is not a suitable host for the *Verticillium* disease organism, and when its residues are incorporated into the soil, they release biofumigants that effectively reduce the presence of various disease-causing organisms. Other non-host crops, such as spinach, peas, and artichokes, have also been successfully integrated into crop rotation systems with strawberries, contributing to disease management and overall soil health improvement.

The Western tarnished plant bug (*Lygus Hesperus*), being a generalist pest, poses significant challenges for control through traditional input substitution methods. However, innovative strategies have been developed to address this issue. In strawberry fields, replacing every 25th row with a row of alfalfa (constituting approximately 3% of the field area) has proven effective. By concentrating control measures, such as vacuuming or the application of biopesticides, on these alfalfa rows, it has been possible to reduce the damage caused by *Lygus* bugs to acceptable levels. Furthermore, alfalfa rows have demonstrated their value as habitats for beneficial insects, facilitating enhanced natural pest control and contributing to the overall balance of the agricultural ecosystem.

Fig. 5.1 Purple perilla as a pushing plant to protect the main crop soybean. (Courtesy: Prof. Luo Shiming, personal bestowing)

There two examples of push-pull technology are being used recently in China. A 'pull' example from soybean field uses the allelopathic plant purple perilla (*Perilla frutescens* L.) as a pushing crop to resist insect of soybean, thus preventing pest from damaging soybean and finally decreasing pesticide use (Fig. 5.1).

A pulling example from rice field in Jianyang County of Fujian, China (Fig. 5.2), in the scene, an insect attractive flower plant was planted at the ridge of rice field, which provides the harbouring habitat for rice pests and diseases, thus decreases pest and disease transmission and finally decreases pesticide use.

In the transformative process, the ever-emerging questions and barriers such as soil-borne diseases leading to root rot urged the agroecological research collaboration between the Cruhan farm and the USSC agroecology program on integrated pest management. After 3 years of efforts on the transition, noticeable changes have been observed in the Cruhan farm, organic agriculture has extended to other regions of California and stimulated the local government to issue the Organic Farming Act.

- Agroecology as a practice refers to using the principles to design a new agricultural model called 'Ecological Agriculture' to promote the agriculture and food system transformation (de Schutter 2011a, b; Gliessman 2011b). In Olivier de Schutter 's report 'Agroecology and the Right to Food', he speculated that agroecology has two parallel meanings, both as science and practice(de Schutter 2011a, b), but it is better to explore 'Ecological Agriculture' representing the 'practice' aspect of agroecology. 'Ecological Agriculture' as a concept was first coined by an American soil scientist, William A. Albrecht in 1970 (Mercier 1978), rightly meaning that integrating ecology into agriculture development, 11 years after, a British agronomist, Martha Kiley-Worthington defined Ecological Agriculture as (Kiley-Worthington 1981):

Fig. 5.2 An example of a pulling crop to protect the main crop in the rice field of Fujian province of China. (Photo credit: Songliang Wang)

> The establishment and maintenance of an ecologically self-sustaining low input, economically viable, small farming system managed to maximize production without causing large or long-term changes to the environment, or being ethically or aesthetically unacceptable.

Certainly, Kiley-Worthington's definition is based on the intensive-input and high-yield conventional agriculture and took Ecological Agriculture as one of the alternative agricultures emerging since the early twentieth century in developed countries. It may not match the status and demands of developing countries whose agricultural productivity stays at a much lower level and whose population remains at a higher increasing speed. How did Ecological Agriculture look like in the latter? China, Romania, and Turkey and other countries took the lead in trying to configure and practice Ecological Agriculture, aiming at maintaining yield per capita without degrading the resource and environment, closer to farming with less harm as the compromising definition of Ecological Agriculture in those countries. In 2009, Wang and Chen proposed a definition of Ecological Agriculture after the long-term practice of Ecological Agriculture in China:

> An agricultural system that combines modern scientific achievements with the essence of traditional agricultural technology by applying the principles of Agroecology and systematic scientific methods in Ecology to achieve functional and ecological rationality.

Under this philosophy, Ecological Agriculture has been practised in those countries particularly in China since the 1980s. Still, it has not been mainstreamed under the prevailing 'production' and industrial-equipped ideology of the agricultural system, even though many successful cases were achieved in those countries (see Box 5.2). When the world faces a serious food security problem and thus agroecological transformation becomes inevitable, Ecological Agriculture will reoccupy the niche of future models towards the sustainable food system rebuilding (see Chap. 10 for more details).

Box 5.2 The Evolution of China's Ecological Agriculture (Wang 2019; Wang and Shi 2023)

China's Ecological Agriculture (CEA) has come a long way and can be divided into two main periods: traditional ecological agriculture and modern ecological agriculture. Traditional ecological agriculture is about the ecological elements that were part of China's traditional agriculture before 1980.

Modern ecological agriculture started in 1981. That's when the concepts of Ecology and Agroecology were brought into China's academic field. Since then, it has gone through four different development stages. First was the initial exploration stage from 1981 to 1992. Then came the expansion and consolidation stage from 1993 to 2004. After that, there was the steady advancement stage from 2005 to 2011. And finally, the innovation and promotion stage which started in 2012 and is still going on today.

Back in 1993, at the start of the consolidation stage of CEA, the government got involved. They set up a five-level demonstration and driving system for ecological agriculture. This system included two pilot provinces for eco-agriculture, 10 demonstration cities for circular agriculture, 100 demonstration counties for eco-agriculture, 12 demonstration bases for eco-agriculture, and 1100 beautiful villages. By 2012, this system covered different levels like provinces, municipalities, counties, bases, and villages.

In 2012, China's Ecological Agriculture became part of the National Strategy of Ecological Civilization. And in 2017, it was included in the National Strategy of Rural Vitalization. Since then, CEA has been an integral part of ecological construction of rural industry that aims at recovering soil ecosystem and producing safe food. To follow the principle of good product, good price of economics, before the market recognition of CEA products, local agroecological compensation policies are built and applied, more important trend is that a total new agricultural organization system named community supported agriculture (CSA) was popularized by educating consumers to prepay CEA famers.

- Agroecology, as a social movement, refers to the agriculture and food system transformation needing a movement of perspective, vertically from local, regional to global, and from top to down within a country; horizontally from production (supply and producer) to consumption (demand and consumer). This in turn requires agroecology to scale up and out simultaneously by expanding from interdisciplinary and transdisciplinary theoretically (Wang et al. 2019) and from a science to a practice, from a practice to a social movement (Wezel et al. 2009). The conclusion of agroecology as a social movement was demonstrated by a group of cases of agri-food system rebuilding in southern hemisphere countries like Cuba, Brazil, and Mexico. Consider Cuba as an outstanding example (IPES-Food 2018; HLPE 2019). Cuba turned to the Western industrial agriculture in 1960 fully embracing the Green Revolution with intensive input aid from socialist countries and energetically using monoculture with heavy agrochemical input to reach the highest yield per capita. However, the yield outcome appeared to decline at the end of 1980s due to soil degradation and other natural resource depletion, Cuban agri-food system seemed untenable after the loss of support from other socialist countries and subsequently the state-owned agricultural farms declined and went bankrupt, gradually. The small-scale farms struggled. In response to the cries, the agroecologists moved their steps from universities and colleges to help the small-scale farmers making the transition, referring to the movement as the 'agroecological revolution' replacing the 'green revolution' that had caused the crisis. Later many NGOs (such as *campesino-a-campesino*, meaning farmer-to-farmer; the Asociación Nacional de Agricultores Pequeños, ANAP or National Association of Small Farmers, facilitated by the agroecologists) were involved in the movement with a banner of rebuilding the nation's agri-food system. As result, the small-scale farms gradually recovered the yield per capita while decreasing agrochemical input to almost half of the previous levels. By the end of 2016, 300,000 small-scale farms joined the agroecology movement, covering 60% of vegetables, maize, fruit, bean, and pork production, and urban agriculture is burgeoning with the same agroecology model (Global Alliance for the Future of Food 2021). The movement was described as follows (IPES-Food 2018):

> Organizing horizontal exchanges between farmers in the field through farmer-to-farmer experience sharing and systematic knowledge exchange; Making the farmers the 'experts' (through several methods of interaction); Recognizing the need to adapt to local conditions (in the development of crop varieties and biological products); Building institutional cooperation between stakeholders in the process.

4 Food Production and Environmental Degradation: Trade-off with Technologies

Technology needs to be considered under any strategy for rebuilding the world's agri-food system. Traditionally, expanding land and enhancing technology are two paths to reaching the goal of crop and animal productivity. However, arable land has been an extremely restrictive factor for reaching the production goal in 2050; but what kind of technology can be of use to reach our goals?

In 1995, Dr. A.S. Ganguly, director of the Institute of Food Research Norwich Laboratory proposed the principles of technology application in the coming future (Ganguly 1995):

Listen carefully to our consumers;
Provide perceivable value for money;
'Naturalness' is a growing requirement;
Sustaining our environment is important

Under those principles, Ganguly made his predictions of the future agri-food technologies:

- Food industries demonstrating sensitivity to the dynamic requirements of consumers are highly likely to attain success. In contrast, for instance, those entities that inadequately manage the information (gathered from consumers) will be confronted with the imperative to either effectuate changes or face the prospect of business decline.
- Biotechnology is set to be the cornerstone of progress in the food industry, yet this hinges crucially on consumer understanding and acceptance. In the agricultural sector, a prime focus will be on aligning production with processing compatibility and quality imperatives. This will be accomplished mainly through two key approaches: accelerated breeding, which hastens the development of desired traits, and direct gene manipulation, allowing for precise modification of genetic material. These strategies in agriculture are integral to leveraging biotechnology's potential in revolutionizing the food industry.
- Chemical additives and processes are anticipated to be supplanted by alternatives derived from natural sources. For instance, novel preservation processes and materials hold the potential to expand existing markets, enhance product quality, and restructure the nature of the distribution chain. In the context of the evolving food and related industries, the shift towards natural alternatives is driven by growing consumer preferences for products with fewer synthetic components, as well as concerns over the environmental and health implications of traditional chemical- based methods. These new preservation modalities may involve the use of natural antimicrobial agents, biopolymers for packaging, and non-chemical-reliant techniques for maintaining product integrity. Such advancements not only offer the promise of improved product characteristics but also signify a significant transformation in the operational and market dynamics of the sectors involved.
- Scientists will be obligated to communicate the extant facts regarding the environmental impact of new technologies in an intelligible fashion to both legislators and the general populace. This necessitates a clear and systematic approach, as the information conveyed must be accessible to a diverse audience. By presenting data in a comprehensive yet understandable way, scientists can facilitate informed decision-making among legislators in the formulation of relevant policies. Simultaneously, for the general public, such communication promotes environmental awareness and understanding, enabling them to engage more effectively with the implications of new technologies in their daily lives.

Therefore, those who thought we could copy technologies from the past are bound to fail, as Einstein said, 'we could not use the ideas to solve the problems that were created by the ideas'. Besides it is impossible to use land expansion for a 50%

grain food production increase in 2050, increasing the intensive external input (the so-called sustainable intensification advocated in the European community, see Pretty and Bharucha 2014, Pretty et al. 2018, etc.) is also needed to trade-off. In industrial countries, both biological and land production potentials are extremely developed. Simultaneously, land degradation and social issues have been extremely serious, meaning lowering the external fossil input and maintaining the resource and environment integrity should be their priority. This needs a system change of thinking and technology of agri-food system. However, in some developing countries, like most countries in Africa and Asia, they need to increase R&D investment to facilitate the high- and neo-technologies to reach the production potentials, without compromising the environment and resources depletion. Success is bound to follow the route of agroecological transformation under the ten elements of agroecology (FAO 2018; Gliessman 2019a, b); the potential technologies for building a sustainable agri-food system are suggested as follows:

- **Biotechnology:** One of the parts of the planet that has the most potential for sustainable improvement agriculture is Africa. African countries need to catch up with newly advanced technologies that could enhance productivity, without environmental degradation, under the stresses of climate change effects such as drought. Biotechnology is a promising technology among the neo-technologies (Juma 2011). Analysts estimated that green revolution that prevailed in 1960–1970s prevented 42 million children from malnutrition in the developing countries, under the current situation, Africa needs the second real Green Revolution and biotechnology will be core in such a new Green Revolution. However, due to biosecurity consideration, African countries are precautionary on genetically modified organisms and have very strict regulation; among the 29 countries that are open to GMO crops, only three are in Africa (Juma 2011). In this regard, African countries need to put this on the discussion table and make a substantial change. Besides GMO technology, gene editing is a newly developed biotechnology that shows great potential in future agri-food system rebuilding. A crop breeding strategy called targeted breeding can be achieved in a shorter timeframe to provide improved traits in crop varieties. Recently, for instance, the CRISPR/Cas9 gene editing technique has been shown to be highly efficient, specific, simple, and cost-effective method for targeted mutations in plants used in Ethiopia. This technology can help develop more tolerant crops to drought, pathogens, and pests, while also enhancing their nutritional value and extending their shelf life, thereby reducing pre- and post-harvest yield losses. Furthermore, by utilizing CRISPR/Cas9, crops can be tailored to better adapt to local conditions, such as extreme temperatures and high altitudes, resulting in higher yields and improved crop quality (de Wit 2022; Marshall 2024). To entirely harness the capabilities of CRISPR/Cas9 for crop improvement, African countries need to invest in research and development in this field. By doing so, the countries can ensure that it benefits from this technology, contributing to the growth and sustainability of its agricultural sector.

- **Information Technology:** The quick development of computer science and information technology allows agri-food system practitioners tackle the messages between society and nature, human and organism, and organism and environment concisely, benefiting future sustainability of agriculture (Schmoldt 2001). By doing so, (1) the resource utilization efficiency will therefore be increased; (2) the environmental pollution could be reduced in developed countries; and (3) value-added can be realized or capitalized in for small-scale farmers by using e-commerce channels (Wang 2005). In the past two decades, an agriculture model named Precision Agriculture supported by Global Position System, Remote Sensing, and Geographic Information Systems on certain traditional advanced machines has provided the novel trade-off solutions between the higher yield and environmental maintenance and resource utilization. Confronting climate change, an agriculture model called climate-smart agriculture has been proposed and practised for mitigating and adapting to the impact of climate change on agri-food systems (FAO 2011; HLPE 2020). All the abovementioned models of agriculture are based on the progresses of computer science and informatics. Recent development in artificial intellectual techniques is offering infinite possibilities for agri-food agroecological transformation as well.
- **Aquaculture Technology:** As mentioned earlier, meat consumption is increasing with increased income in developing countries, but the competition between food and feed has exacerbated the effects of grain production on the environment and GHGs emissions globally. Therefore, turning to aquatic systems for protein food is a promising option. Traditionally, people have harvested protein food from the seas and rivers by way of direct capture; unfortunately, humans have exhausted the fish resource and global wild catches are decreasing. Aquaculture becomes a choice in the new era. By doing so, we will be able to convert vegetative biomass to animal protein with higher efficiency if we select aquaculture species low in the food chain like carp. There is also an opportunity to combine agriculture and aquaculture in the same time and same space such as in cultivated fish in paddy fields. This is a possibility for harvesting animal protein food in water-rich developing countries. Meanwhile, practitioners and managers need to think about bio-safety and environment pollution issues while running an aquaculture farm.

References

Berry, Wendal. 1977. *The unsetting of America: culture and agriculture*. San Francisco, CA: Sierra Club Books.

Biovision Foundation for Ecological Development & IPES-Food. 2020. *Money flows: what is holding back investment in agroecological research for Africa?* Biovision Foundation for Ecological Development & International Panel of Experts on Sustainable Food Systems. www.ipes-food.org. Accessed March 19, 2025.

Brown, Lester R. 1995. *Who will feed China? Wake-up call for a small planet*, 163. New York: W.W. Norton & Company, Inc. Worldwatch Institute, Environmental Alert Series.

Crist, Eileen, Camilo Mora, and Robert Engelman. 2017. The interaction of human population, food production, and biodiversity protection. *Science* 356 (6335): 260–264. https://doi.org/10.1126/science.aal2011.

Dale, Bryan. 2020. Alliances for agroecology: from climate change to food system change. *Agroecology and Sustainable Food Systems* 44 (5): 629–652.

Dale, Bryan. 2023. Food sovereignty and agroecology praxis in a capitalist setting: the need for a radical pedagogy. *Journal of Peasant Studies* 50 (3): 851–878.

de Schutter, Olivier. 2011a. *Agroecology and right to food.* Report presented at the 16th Session of the United Nations Human Rights Council[R/OL]. http://www.srfood.org/images/stories/pdf/officialreports/20110308_a-hrc-16-49_agroecology_en.pdf. Accessed June 17, 2024 in Canada.

de Schutter, Olivier. 2011b. *Eco-agriculture (agroecology) can double food production in 10 years.* http://www2.ohchr.org/english/issues/food/index.htm (cited from Stephen R. Gliessman 2011. Editorial: Agroecology and the Growing Food Crisis. J Sustain Agric. 35:697–698).

de Schutter, O. 2012. Agroecology, a tool for the realization of the right to food. In *Agroecology and strategies for climate change*, Sustainable agriculture reviews, ed. E. Lichtfouse, vol. 8. Dordrecht: Springer. https://doi.org/10.1007/978-94-007-1905-7_1.

de Schutter, Olivier. 2014. *Agroecology and the right to food.* Interview by Margriet Goris. http://leisaindia.org/wp-content/uploads/2014/07/LEISA-India-June-2014-Page-18-19.pdf. Accessed March 24, 2025.

de Wit, Maywa Montenegro. 2022. Can agroecology and CRISPR mix? The politics of complementarity and moving toward technology sovereignty. *Agriculture and Human Values* 39:733–755.

Deaconu, Ana, Geneviève Mercille Ekomer, and Malek Batal. 2021. Promoting traditional foods for human and environmental health: lessons from agroecology and Indigenous communities in Ecuador. *BMC Nutrition* 7:1.

FAO (Food and Agriculture Organization of the United Nations). 1996. *Rome Declaration on World Food Security and World Food Summit Plan of Action.* https://www.fao.org/3/W3613E/W3613E00.htm. Accessed July 6, 2024.

FAO (Food and Agriculture Organization of the United Nations). 2011. *Climate-smart agriculture: managing ecosystems for sustainable livelihoods.* Rome: Food and Agriculture Organization, Unite Nations.

FAO (Food and Agriculture Organization of the United Nations). 2018. *The 10 elements of agroecology: guiding the transition to sustainable food and agricultural systems.* Rome: FAO.

FAO (Food and Agriculture Organization of the United Nations). 2021. *The State of Food Security and Nutrition in the World 2021: The world is at a critical juncture.* https://www.fao.org/state-of-food-security-nutrition/2021/en/. Accessed April 27, 2024.

Ferguson, Bruce G., Helda Morales, Kimberly Chung, and Ron Nigh. 2019. Scaling out agroecology from the school garden: the importance of culture, food, and place. *Agroecology and Sustainable Food Systems* 43 (7–8): 724–743. https://doi.org/10.1080/21683565.2019.1591565.

Francis, C., G. Lieblein, S. Gliessman, T. A. Breland, N. Creamer, R. Harwood, and R. Salvador. 2003. Agroecology: the ecology of food systems. *Journal of Sustainable Agriculture* 22 (3): 99–118.

Ganguly, A. S. 1995. Science of food in the twenty-first century. *Nutrition & Food Science* 2:17–20.

Gliessman, Stephen R., ed. 2003. *Agroecology: the ecology of sustainable food systems.* Boca Raton, FL: Taylor and Francis.

Gliessman, Stephen R. 2011a. Agroecology and food system change. *Journal of Sustainable Agriculture* 35 (4): 347–349.

Gliessman, Stephen R. 2011b. Editorial: agroecology and the growing food crisis. *Journal of Sustainable Agriculture* 35 (7): 697–698.

Gliessman, Stephen R. 2019a. Feeding prometheus: agroecology and unchaining our desire for food system transformation. *Agroecology and Sustainable Food Systems* 43 (1): 1–2. https://doi.org/10.1080/21683565.2018.1515541.

Gliessman, Steve R. 2019b. Making the transition away from industrial agriculture: using agroecology to promote food systems transformation. *Agroecology and Sustainable Food Systems* 43 (2): 121–122. https://doi.org/10.1080/21683565.2018.1529358.

Gliessman, Stephen R. 2020. Transforming food and agriculture systems with agroecology. *Agriculture and Human Values* 37:547–548.

Global Alliance for the Future of Food. 2021. *The politics of knowledge: understanding the evidence for agroecology, regenerative approaches, and indigenous foodways*. n.p.: Global Alliance for the Future of Food.

HLPE (the High-Level Panel of Experts). 2019. *Agroecological and other innovative approaches for sustainable agriculture and food systems that enhance food security and nutrition*. A report by the High Level Panel of Experts on Food Security and Nutrition of the Committee on World Food Security, Rome, Italy.

HLPE (the High-Level Panel of Experts). 2020. *Food security and nutrition: building a global narrative towards 2030*. A report by the High Level Panel of Experts on Food Security and Nutrition of the Committee on World Food Security, Rome.

IPES-Food. 2015. *The new science of sustainable food systems: overcoming barriers to food systems reform*. First Report of the International Panel of Experts on Sustainable Food Systems.

IPES-Food. 2016. *From uniformity to diversity: a paradigm shift from industrial agriculture to diversified agroecological systems*. International Panel of Experts on Sustainable Food Systems. www.ipes-food.org. Accessed July 5, 2024.

IPES-Food. 2017. *Too big to feed: exploring the impacts of mega-mergers, concentration, concentration of power in the agri-food sector*. www.ipes-food.org. Accessed March 19, 2025.

IPES-Food. 2018. *Breaking away from industrial food and farming systems: seven case studies of agroecological transition*. www.ipes-food.org. Accessed July 5, 2024.

Jackson, Dana L., and Laura L. Jackson Johnson. 2002. *The farm as natural habitat: reconnecting food systems with ecosystems*, 207. Washington. DC: Island Press.

Juma C. 2011. Preventing hunger: Biotechnology is key. *Nature*, 479: 471–472

Khan, Z., C. Midega, J. Pittchar, J. Pickett, and T. Bruce. 2011. Push–pull technology: a conservation agriculture approach for integrated management of insect pests, weeds and soil health in Africa. *International Journal of Agricultural Sustainability* 9:162–170.

Kiley-Worthington, M. 1981. Ecological agriculture: what it is and how it works. *Agriculture and Environment* 6 (4): 349–381.

Kirschenmann, Frederick L. 2007. Potential for a new generation of biodiversity in agroecosystems of the future. *Agronomy Journal* 99:373–376.

Marshall, Michael. 2024. Creating new crops with CRISPR. *Nature* 625:230–232.

Martínez-Torres, M. E., and P. M. Rosset. 2010. La Vía Campesina: the birth and evolution of a transnational social movement. *Journal of Peasant Studies* 37 (1): 149–175.

Mercier, J. R. 1978. Towards ecological agriculture. *Futuribles* 16:479–484.

Morgan, Caitlin B., and Amy B. Trubek. 2020. Not yet at the table: the absence of food culture and tradition in agroecology literature. *Elementa Science of the Anthropocene* 8:40. https://doi.org/10.1525/elementa.437.

Muramoto, J., S. R. Gliessman, S. T. Koike, C. Shennan, C. T. Bull, K. Klonsky, and S. Swezey. 2014. Integrated biological and cultural practices can reduce crop rotation period of organic strawberries. *Agroecology and Sustainable Food Systems* 38:603–631.

Pretty, Jules. 2002. *Agri-culture: reconnecting people, land and nature*. London: Earthscan Publications Limited.

Pretty, Jules. 2003. Agroecology in developing countries: the promise of a sustainable harvest. *Environment: Science and Policy for Sustainable Development* 45 (9): 8–20.

Pretty, Jules, and Zareen Pervez Bharucha. 2014. Sustainable intensification in agricultural systems. *Annals of Botany* 114 (8): 1571–1596.

Pretty, Jules, Tim G. Benton, Zareen Pervez Bharucha, Lynn V. Dicks, Cornelia Butler Flora, H. Charles, J. Godfray, Dave Goulson, Sue Hartley, Nic Lampkin, Carol Morris, Gary Pierzynski, P. V. Vara Prasad, John Reganold, Johan Rockström, Pete Smith, Peter Thorne, and Steve Wratten. 2018. Global assessment of agricultural system redesign for sustainable intensification. *Nature Sustainability* 1:441–446.

Rahman, Diana, Theano Moussouri, and Georgios Alexopoulos. 2021. The social ecology of food: where agroecology and heritage meet. *Sustainability* 13:13981. https://doi.org/10.3390/su132413981.

Ramankutty, Navin, and Hadi Dowlatabadi. 2021. Beyond productivism versus agroecology: lessons for sustainable food systems from Lovins' soft path energy policies. *Environmental Research Letters* 16:091003.

Rosset, Peter M. 2011. Preventing hunger: change economic policy. *Nature* 479:472–473.

Schmoldt, D. L. 2001. Editorial: Precision agriculture and information technology. *Computers and Electronics in Agriculture* 30:5–7.

Springmann, Marco, Michael Clark, Daniel Mason-D'Croz, Keith Wiebe, Benjamin Leon Bodirsky, Luis Lassaletta, Wim de Vries, Sonja J. Vermeulen, Mario Herrero, Kimberly M. Carlson, Malin Jonell, Max Troell, Fabrice DeClerck, Line J. Gordon, Rami Zurayk, Peter Scarborough, Mike Rayner, Brent Loken, Jess Fanzo, H. Charles J. Godfray, David Tilman, Johan Rockström, and Walter Willett. 2018. Options for keeping the food system within environmental limits. *Natue* 562 (7728): 519–524.

Swezey, S. L., D. J. Nieto, J. R. Hagler, C. H. Pickett, J. A. Bryer, and S. A. Machtley. 2013. Dispersion, distribution, and movement of Lygus spp. (Hemiptera: Miridae) in trap-cropped organic strawberries. *Environmental Entomology* 42:770–778.

Tilman, David, Joseph Fargione, Brian Wolff, Carla D'Antonio, Andrew Dobson, Robert Howarth, David Schindler, William H. Schlesinger, Daniel Simberloff, and Deborah Swackhamer. 2001. Forecasting agriculturally driven global environmental change. *Science* 5515 (292): 281–284.

United Nation, Department of Economic and Social Affairs Population Division. 2024. World Population Prospects. https://population.un.org/wpp/. Accessed Dec.3. ,2025

Vandermeer, John. 2011. *The ecology of agroecosystems*, 392. Sudbury, MA: Jones and Bartlett Publishers.

Vandermeer, John, Aniket Aga, Jake Allgeier, Catherine Badgley, Regina Baucom, Jennifer Blesh, Lilly F. Shapiro, Andrew D. Jones, Lesli Hoey, Meha Jain, Ivette Perfecto, and Mark L. Wilson. 2018. Feeding prometheus: an interdisciplinary approach for solving the global food crisis. *Frontiers in Sustainable Food System* 2:39. https://doi.org/10.3389/fsufs.2018.00039.

Vermeulen, Sonja J., Bruce M. Campbell, and John S. I. Ingram. 2012. Climate change and food systems. *Annual Review of Environment and Resources* 37:195–222.

Wang S. 2005. ITs: Toward the sustainable Agroecosystem Management(in Chinese). *Agricultural Network Information*, 8:4–7,12

Wang, S. L. 2019. Concerted development of ecological agriculture along with community-supported agriculture to facilitate rural vitalization. *Chinese Journal of Eco-Agriculture* 27 (2): 212–217.

Wang, S. L., and D. M. Chen. 2009. Development of modern ecological agriculture in Fujian: achievements, problems, and countermeasures (in Chinese). *Journal of Fujian Agriculture and Forestry University (Philosophy and Social Sciences)* 12 (4): 25–29.

Wang, S. L., and S. X. Shi. 2023. Developing Chinese ecological agriculture acts as a vital pathway of the state-specifically agricultural modernization in China: a historical review of ecological agriculture movement in China. *Chinese Journal of Eco-Agriculture* 31 (8): 1184–1193.

Wang, S. L., C. D. Caldwell, S. L. Kilyanek, and Sean M. Smukler. 2019. Using agroecology to stimulate the greening of agriculture in China: a reflection on 15 years of teaching and curriculum development. *International Journal of Agricultural Sustainability* 17 (4): 298–311.

Wezel, A., and C. David. 2012. Agroecology and the food system. In *Agroecology and strategies for climate change*, Sustainable agriculture reviews, ed. E. Lichtfouse, vol. 8. Dordrecht: Springer. https://doi.org/10.1007/978-94-007-1905-7_1.

Wezel, A., S. Bellon, T. Doré, C. Francis, D. Vallod, and C. David. 2009. Agroecology as a science, a movement and a practice. A review. *Agronomy for Sustainable Development* 29 (4): 503–515.

William, Justine M. 2012. *agroecology versus productivism: completing discourses on the future of Cuban agriculture*. Chapel Hill, NC: University of North Carolina.

Chapter 6
Integrating Rural Landscape Sustainability into Agroecology

Abstract Currently, the study of agroecology does not include a significant discussion of rural development which is closely linked with agri-food systems. This chapter addresses the principally agriculture-based natural and social landscapes in rural areas. Agroecology, defined as a science of the whole food system would benefit from the incorporation of ecology of rural landscapes to address the complexity of the rural ecosystems in a better way in which agriculture is the dominant human intervention. In developing countries, there is a comparatively strong dichotomy of life experience between rural and urban landscapes. The intent of including rural ecology in the larger discipline of agroecology is to provide models and approaches that can guide a process of rebuilding rural ecosystems.

Keywords Sustainable development · Rural landscape sustainability · Rural and urban dividend · Rural ecology · Rural China · Agroecology

Learning Objectives

After reading this chapter, students should be able to:

1. Characterize the causes and impacts of the dichotomy of experience between rural and urban areas in developing nations.
2. Describe the close links between rural landscape and agriculture.
3. Explain the potential roles that agroecology can play in building rural sustainability.
4. Describe the proposed discipline of rural ecology in both its disciplinarity and scale.
5. Discuss the role of rural ecology in building a sustainable rural ecosystem.

****

> There are 2 spiritual dangers in not owning a farm, one is the danger of supposing that breakfast comes from the grocery, and the other is that heat comes from the furnace.
>
> —Aldo Leopold (1949) (Cited from Dana L. Jackson and Laura L. Jackson 2002)

S. Wang, C. D. Caldwell, *Advanced Agroecology*,
https://doi.org/10.1007/978-3-032-08689-1_6

> If we don't work hard and work together to improve farmers' profitability and secure a land base for farming in the face of urban development, many of us will live in sprawl, drive through sprawl and work in sprawl.
> —Dana L. Jackson and Laura L. Jackson (2002)

The modern crisis of the agri-food system, which is characterized by issues such as human health risks, environmental degradation, biodiversity loss, and overproduction tied with farmer poverty is embedded in the mode of rural development imposed by the effects of market-oriented economics and production-biased agriculture (Marsden et al. 2001; Jarosz 2012). Since the agri-food system is profit centred, and the basis for production (farmers) is undervalued in the value chain, the primary production sector and the rural landscape have not had the resources for economic or social sustainability. The international free market under the guise of globalization has resulted in rural dilapidation.

To find a sustainable, equitable path forward, it is necessary to reflect and critically evaluate the reasons for rural decline. Agroecology (since it encompasses both a science and social movement) provides a clear logical direction, considering human nature, the nature of society, and human-human relationships for reconstructing the rural ecosystem. In the context of 'sustainable livelihoods' for rural sustainability, agroecology could be considered as two disciplinary branches of production agroecology and political agroecology (Amekawa 2011; Tilzey 2024), the combination of which refer to using radical food democracy ideology to define food as public good rather than a capital commodity (Tornaghi and Dehaene 2020; Tilzey 2024). Those two aspects of agroecology could be integrated into rural ecology, a newly emerged citizen science (Wang et al. 2021, Wang and Caldwell 2016).

1 Rural Dilemma in Developing Nations

The urgent problems or challenges in rural areas, particularly in developing countries, are economic, social, and environmental interwoven in a multisectoral function (Ellis and Biggs 2001; Jung and Yoo 2023). The interacting factors can be summarized as follows:

Rural PovertyA large proportion of the world's population lives in developing countries. In 2024, 8.5% of the world's population was living in extreme poverty. Millions of people are still trapped in poverty despite national country progress under globalization and free trade (World Bank 2024). Such poverty is related to a lack of education, unemployment, and depleted or spoiled environmental resources. Currently, 3.45 billion of the world's citizens live in rural areas (United Nations 2024; World Bank 2024).

- Food Insecurity
  Population pressure has posed difficulties for rational use of land and maintenance of the rural natural environment; food-producing activities have degraded land and decreased land productivity, resulting in food insecurity. In many

regions of the world, the number of hungry people has continued to grow until, as of 2023, the number of undernourished people has reached 757 million due to overuse, ill-use of the land and environmental resources, and other economic and political problems in rural areas (FAO 2024).

- Environmental and Resource Degradation
  A larger national population now relies on much less rural cropland than in the past, making it even more important to oppose practices which cause land degradation, biodiversity loss, genetic erosion of crop germplasm, and environmental (soil, air, and water) deterioration. Since the Green Revolution, grain production has relied heavily on a narrow germplasm base (Brush 1989; Thomas and Kevan 1993; Julia-Serres et al. 2019), an overuse of synthetic fertilizers and pesticides has resulted in soil degradation, river eutrophication, environment contamination, and food residues that have produced unanticipated food safety problems (Hooker 1999). Rural environments in developing countries have suffered profoundly from policies aimed at increasing returns and economic revenues to support the increasing national population (Midmore and Whittaker 2000; Landa et al. 1997).

Such problems are not the result of economic, ecological, or social processes individually, but rather stem from the cumulative interaction of those processes. To address these interwoven processes, a broader, cross-discipline approach with holistic, hierarchical views and ecological perspectives is required. The goal is to harmonize the above threefold challenges to define and develop a rural renaissance of wholeness and complexity (Altieri 1998; Marsden et al. 2001; Vasishth 2008).

Unfortunately, at present, there is no discipline dealing with such complex situations in the magnitude experienced in rural areas of developing countries. The science of rural people tends to be ignored or separated from the larger context of daily life, labour, and livelihood (Rocheleau 1991); social science and politics tend to downplay agriculture as a central place for rural livelihood (Marsden et al. 2001). Rural sociology and agricultural economics are two disciplines involved in the study of rural life, but both have not been successful in grasping natural dimensions of rural relations (Marsden et al. 2001; Goodman 1999). Economics tends to deal with the abstractions of optimal allocation of resource using marketing power. That pursues the maximal economic benefit but is less sensitive to sociological aspects. It often ignores the fact that land resource in rural areas is not only a production factor but also the farmers' home and life resource. With market changes, land resources cannot be reallocated solely by market power (Wen 2001). Ecological and socio-cultural settings in rural areas are very complex; therefore, multi-, inter-, and cross-disciplinary science and systematic approaches are needed to deal with such fields and many agriculture-orientated problems in rural landscapes which differ greatly from the urban area (Field et al. 2003).

It is crucial to incorporate ecological principles and knowledge into the study of the complex processes of rural landscape and societal relations as a holistic system (Hayami 2001). As Goodman (1999) stated, 'ecological' should become a central theme of rural and agri-food studies. Presently, general ecology favours ideology that focuses on the understanding and emphasis on organisms, partly distorted from its original definition that deals with the relationship between living beings and their environment (Tricart and Kiewietde 1992). What is needed is to use more modern approaches to ecology which will investigate the interactions of various elements of the co-evolution of living beings (agronomic organism and humans in this case) in rural areas (Marsden et al. 2001).

Speaks of ecological modernization towards new rural eco-realties that can make rural development sustainable. Agroecology offers a much clearer direction to the analysis of the relationship of humans with nature in rural development (Marsden et al. 2001). A United Nations Special Rapporteur on the Right to Food also insisted that the pathway to solve issues like food sovereignty and insecurity is via an agroecological approach (Méndez et al. 2013; De Schutter 2011). Therefore, this chapter supports rural ecology as an emerging sub-discipline of agroecology in developing countries including China to promote more holistic analyses and approaches to rural development. In the future, this must go beyond the narrow boundaries of traditional agricultural disciplines and education for rural development. Internationally, this disciplinary approach is new both in literature and in practice. The situation in developed countries is different from that in developing countries since the proportion of rural areas is different and agricultural economies have basic differences. This chapter is an overview and characterization of rural ecology, followed by a proposal for an analytical framework of rural ecology for developing countries.

The major challenges are to synthesize and apply this knowledge in rural development within the scope of rural ecology. In addition, activities must address the problems of rural communities in their true complexity and in a holistic way that relates to the development of the individual agricultural industry, food production system, and working culture.

2 Integrating Rural Landscape Sustainability into Agroecology

Before exploring the emerging rural ecology for instructing rural sustainable development in more depth, it is necessary to reiterate the roles agroecology plays in the transformation of rural development. They are as follows, aimed at food security, farmer empowerment, and sustainable likelihood of rural areas.

- Agroecology legitimizes sustainable rural development by protecting diversity from the effects of rural development as it is presently expressed in the development strategy of society. Rural development is closely associated with agriculture and food production, but it should be understood that the reductionist agri-food system assigned the production sector to the rural landscape and farmers. Unfortunately, the larger agri-food system has devalued and overwhelmed farmers and rural society. Benefits and profits, for the most part, are taken up by urban industrial groups. The power in the food chain remains in urban hands which is in overall control of resources and pricing. In addition, overproduction in rural landscapes is considered an error by farmers and it is attached to farmer poverty and environmental degradation, The urban mainstream exploits the industrialized agriculture model to devalue rural nature and farmers and deploy a bureaucratic hygienic food model to deprive the sovereignty of farmers (such as crop seeds) and finally scorn the rural areas as 'minefields' and 'graveyards' in the minds of politicians and all urban people, sometimes even the rural residents themselves (Marsden et al. 2001). It is time to legitimize the rural landscape and people in the sustainable map of the whole society. Agroecology is defined as a social movement that will help revalue rural development via its farmer empowerment function (de Schutter 2011); at the least, agroecology offers a process to review and criticize the urban top part of the food chain and its effects on the rural bottom. The narrative is summarized in Box 6.1 by taking European Common Agriculture Policy as a starting case.

Box 6.1 Reforming the Common Agriculture Policy, Exploring Agroecology as a Tool for Legitimizing and Theorizing Rural Development in Europe (van Zanten et al. 2014)

- **Background**
  As a response to the unfavourable consequences that threatened the Agri-food system in Europe due to the postwar adaptation of industrial, chemical-centred agriculture, the 'common agricultural policy (CAP)' was proposed and applied after 1990 to subsidize farmers to reverse the unfavourable practices. However, it seemed such subsidy-based policy has neither changed the capitalized and industrialized agriculture nor has it transformed the food system into an agroecologically compatible one. It appears that an integration of agroecology to CAP reform is imperative in facilitate the European Agri-food system to be sustainable in the new millennium continentwide in rural Europe.
- **CAP restricting processes**
  CAP was established in 1992 to rescue the unsustainable agricultural system based on status analysis of sustainable development in agriculture in Europe before the EU commission was founded. The critical points focused on: (1) production acquired concomitant quality loss and environmental degradation; (2) economic increase without real development in rural area where CAP acted. Shortly after the new millennium, the EU commission responded to the emergence of the broadly accepted concepts of ecosystem services and natural capital as effective instruments to respond to the conflict between economic development and environ mental degradation. The commission issued the reformed CAP which highlighted direct payment of agroecosystem ser vices to farmers, promoting rural development.

(continued)

Box 6.1 (continued)

- **Outcomes**
  - Legitimizing rural development by devolution of the central right of Agri-food system
  - Theorizing agroecological transformation by revaluing and capitalizing the agroecosystem services
  - Increasing the cash income of rural families by realizing agricultural multifunctionalities

- Agroecology initiates an analytical framework of rural sustainable livelihood (SL) by utilizing its social-science aspect (Marsden et al. 2001). The conventional model of rural development has focused mainly on economic and biological production and has lost the connection with sustainable livelihood for rural residents. A sustainable livelihood approach points to the 'good life' of rural households, embodying economic, environmental, and social sustainability, the three pillars of sustainable development. Agroecology is positioned as one of the ethnosciences that link ecology to people's livelihood (Altieri 1993). Therefore, the study of agroecology and rural sustainable livelihood shares not only the full agricultural spectrum but also divergent interests to signify the potential for beneficial cross-fertilization of rural development. This opens innovative arenas in the studies of sustainable agriculture and sustainable rural development that recognize people's good life (Amekawa 2011). The following case was offered by Addinsall et al. (2015, see in Box 6.2).

Box 6.2 The Study Framework of Agroecology and Sustainable Rural Livelihoods in the Pacific Islands

- **Background**
  Several states in the Pacific Islands, such Fiji and Vanuatu, have followed the liberal economic development model without integration of local culture and without consideration of sustainable livelihood of rural people. Therefore, there has been a call to integrate economic and ecological processes towards sustainable livelihood of rural people in the pacific islands. Agroecology defined as a food system approach allowing for the study of interaction and integration between natural and social sciences will take the role appropriately.
- **Agroecology and sustainable rural livelihoods framework**
  Since many development projects in the Pacific Island countries have failed to focus on livelihood of residents, let alone the sustainable 'need' a key word in the concept of sustainable development. A scaling up and out of agroecology will integrate the theory of sustainable development to

(continued)

Box 6.2 (continued)

form the objectives and approaches for rural sustainable livelihood. In such a study framework, the small-scale farmers of Fiji and Vanuatu were put in the Centre, for which the financial, human, physical, social, and natural capitals were collected to service the approach of participatory discussion to reach compromising good livelihood goals, including economic income, food security, ecosystem resilience, equity, and sustainability.

- **Outcomes**
  - Achieving a trade-off strength between the subsistence economy and ecosystem resilience.
  - Achieving the livelihood assets, resources, and capabilities with agro-ecologically transformative smallholders.
  - Providing an avenue to address modern issues such as the need for cash economy without sacrificing the need for long-term sustainability.
  - Developing a sustainable management of agroforestry project in Vanuatu and Fiji.

- Agroecology leverages the agri-food system for public good by integrating its natural and social aspects (FAO 2017). Confronting the problems posed by the current industrialized agriculture model, such as low income and thus food insecurity for agri-food production-based farmer families, and social conflicts by destroying the economic and cultural bases of rural communities, agroecological transformation of the agri-food system might be the only way out of the currently devastated situation of the global agri-food system. In this context, agroecology will fuel such transformation. Agroecology originated from a combination of ecology and agronomy starting in 1928 (Bensin 1928) and has evolved into the ecology of the entire food system (Francis et al. 2003). To achieve the goals of the entire food system, under the current ideology and practice of the agri-food system, agroecology has now been defined as a social movement (Wezel et al. 2009; Rosset and Martínez-Torres 2012). Under this circumstance, rural development is associated with an agri-food system, and agroecology endows such development with a legal standing in seeking harmony among the various dimensions of sustainability for the rural ecosystem. Agroecology, as a science or scientific discipline, aims to transition from the current development model to a more sustainable and equitable one. Agroecology, as a practice, finds ecologically based technologies in place of matter and energy-consuming applications. Agroecology, as a social movement, offers the possibility for extending the new model worldwide (see Chap. 5). Integrating the natural and social scientific aspects of agroecology will pave the way to three-dimensional food system goals simultaneously (Fig. 6.1, adapted from Altieri 1995). Another case was recorded by José Rover et al. (2017, see in Box 6.3).

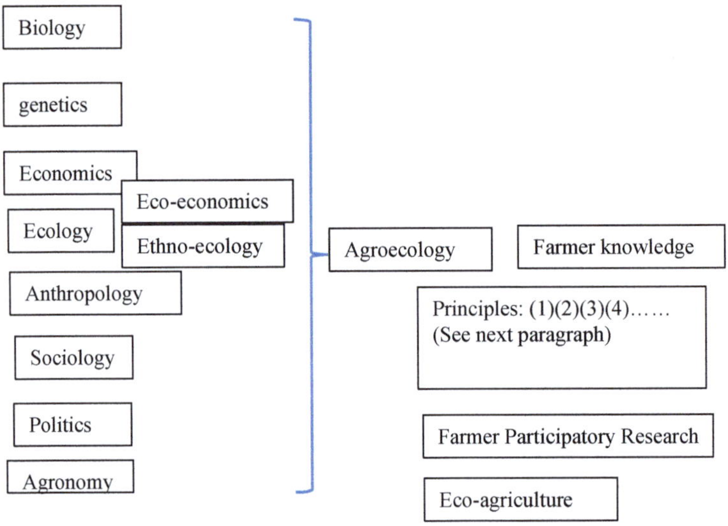

Fig. 6.1 Integrating natural and social sciences in agroecology for agri-food system transformation

Box 6.3 Ecovida Agroecology Network in Brazil: Rebuilding Agri-Food System and Facilitating Rural Development

- **Background**
  As in other places in the world at the end of twentieth century, while increasing a greater amount of food availability, the growing industrialization and capitalization of the Agri-food system in Brazil have led to an increasing risk of food safety and a growing demand from consumers and society for healthy and safe food. Since the top-down policy-driven force for the Agri-food system transformation did not seem to happen, a societal innovation was bound to happen in response to the demand. Ecovida agroecology network was set up as the platform for social innovation for Agri-food and rural development at the time in Brazil.
- **Ecovida Agroecology Network in Brazil**
  Ecovida Agroecology Network, whose original name in Spanish is Rede Ecovida de Agroecologia, was organized in November 1998 in the state of Santa Catarina in 1998 in response to a local government's organic food authentication plan. A group of non-governmental organizations including Farmer Group (FG), the Rede Tecnologias Alternativas (Alternative Technology Network), gathered to discuss how to respond to the government's initiative. The proposition of Ecovida focused on a completely agroecological transformation along the regional rather than the agri-food

(continued)

Box 6.3 (continued)

product labelling and certification strategy, in so doing, the transformation would benefit the agroecological family farmer rather than the agri-food based capital and industrial companies. To reach the goal, Ecovida gradually established the rules of debate to optimize the decision-making process, the famous and significant case of Ecovida is to initiate the Participatory Guarantee Systems (PGS) instead of government-driven certification process in guaranteeing the agri-food products.

- **Outcomes**
  - Provoked innovations that pertain to its horizontal and decentralized setup, the participatory system it employs for certifying organic food, and the energetic relationship it has with the markets based on local trade activities and reciprocal ties.
  - Exerted influence over public sector policies and spurred enhanced cooperation between social actors hailing from rural and urban regions.
  - Initiated the Participatory Guarantee Systems (PGS) to certificate the agroecologically producing products (principle and praxis of PGS are summarized in Chap. 10 associated with the discourse of ecological agriculture and community supported agriculture).

- Agroecology offers a holistic strategy by applying its embedding principles of ecology to manage natural resources sustainably. The viable agriculture and food system was built on a natural ecosystem; therefore, managing natural resources and conserving the environment become prerequisites to forming a sustainable agroecosystem (Altieri 2002). Basically, agroecology offers five principles of ecology to realize the goal (Altieri 1995; Barrios et al. 2020):
  - **'Improving efficiency in the use of resources'** by making the best of the solar radiation, air, water, and nutrients by spatial deployment of agricultural components, such as agroforestry as one model of ecological agriculture.
  - **'Promoting good governance of both natural and human systems'** by recycling the crop biomass and maintaining nutrient balance along the food chain within the agroecosystem, thus minimizing the material and energy external inputs.
  - **'Conserving, protecting, and enhancing natural ecosystems**' by keeping the soil healthy and conserving biodiversity above and beneath soil by using organic 'wastes' within the agroecosystem.
  - **'Enhancing resilience of people, communities, and ecosystems'** via facilitation of integrated pest management. This involves decreasing the use of synthetic pesticides which thereby promotes biodiversity and increases health of the whole food chain.

- **'Protecting and improving rural livelihoods, equity, and social well-being' by** leveraging farmer's income via sound policy on payment for agro-ecosystem services or ecological agriculture products.
  These principles bring agroecology into practice, for example, the Miombo woodlands management in Africa (Box 6.4).

Box 6.4 Sustainable Management of Miombo Woodlands (Gumbo et al. 2018)

- **Background**
  The Miombo woodland is a large African dryland forest ecosystem spanning nearly 2.7 million km^2 across southern African countries such as Angola, the Democratic Republic of Congo, Malawi, Mozambique, Tanzania, Zambia, and Zimbabwe. Dominated by Brachystegia species, either singly or in combination with Julbernardia and Isoberlinia species, it provides numerous goods and services. These include non-wood forest products (like insects, mushrooms, fruits, tubers, medicine, fodder, honey, and seeds) and wood fuels, collectively referred to as non-timber forest products (NTFPs). It sustains the livelihoods of over 100 million rural poor and 50 million urban people, with the charcoal sector employing many rural residents and providing extra income to poor rural families. Communities depend directly on the woodlands for food and nutrition, as NWFPs add essential nutrients to local diets and wood fuel is crucial for cooking and sterilization. However, a growing population in need of agricultural land, unsustainable use, and overharvesting of natural resources in parts of the Miombo woodlands, along with climate change impacts (such as drought and fires), prevent many trees and associated species from regenerating naturally. Additionally, issues like lack of an enabling policy environment, unsustainable management, limited willingness and ability to pay for and access energy-efficiency technologies, inadequate awareness and information (including technical capacity), high poverty levels, and limited access to microcredit facilities compound the problems and impede the development of the Miombo ecosystem. Agriculture is significant in the Miombo woodlands, with crops like maize, groundnuts, beans, and tobacco being grown, which are essential for local livelihoods.
- **Main Strategies**
  - Integrated Silvicultural Systems: Develop and implement integrated silvicultural systems to mitigate disturbances, open the canopy, stimulate soil seed germination, and encourage seedling growth.
  - Sustainable Management: Practice proper management of the woodlands to provide resilience for rural communities, offer nutritious edible products, wood fuel for cooking, and conserve biodiversity and water resources, thus preventing land degradation and desertification.

(continued)

Box 6.4 (continued)

  - Decision-Maker Support: Decision-makers should prioritize conservation and sustainable development, ensuring equitable access to benefits from ecosystem services in the woodlands.
- **Main Outcomes**
  - Livelihood Support: The Miombo Woodlands support the livelihoods of over 300 million people through various non-timber forest products.
  - Ecosystem Conservation: Proper management of the woodlands contributes to resilience for rural communities, conserves biodiversity, buffers extreme weather conditions, and prevents land degradation and desertification.
  - Sustainable Futures: A combination of the strategies can lead to just and sustainable futures in the Miombo Woodlands, benefiting empowered local communities. In 2022, Mozambique led the Miombo Woodlands Declaration, signed by 11 Southern African Development Community countries, which prioritize the sustainable management of the ecosystem's natural resources.

3 An Emerging Rural Ecology as a Subset of Agroecology for Rural Sustainability

Although agroecology is suggested to promote rural development transitioning from a traditional economic model to an ecologically sound developmental model, it is somewhat limited by its six focus on the agricultural sector, which is restrictive in terms of covering the full scale of rural livelihoods (Amekawa 2011). Sustainable livelihoods in rural areas usually go beyond the agricultural processes and products. Therefore, another transdisciplinary approach to ecology, that is able to cover rural livelihoods, is needed to address rural development more completely (Wang et al. 2021).

As addressed in Sect. 1 of this chapter, to fully instruct rural sustainability, the emerging discipline of rural ecology is proposed as a subset of agroecology (Wang et al. 2021; Wang and Caldwell 2016). Haila (1989) stated that any rapidly developing scientific discipline needs to be distinguished by a conceptual framework. Dalgaard et al. (2003) discussed the validity of agroecology as an inter-discipline. It is also the same problem of disciplinarity for rural ecology, is it a multidisciplinary, a pluri-disciplinary, an interdisciplinary, or a transdisciplinary field? Or is it a discipline of its own, or is it becoming one? This section explores the key definitions of the relationship of 'Rural' to 'Ecology', attempting to focus these keywords on the disciplinarity and scales of Rural ecology.

3.1 Disciplinarity of Rural Ecology

An understanding of the differences between natural and social landscape and how they relate to each other is a first step in understanding the disciplinarity and scales of rural ecology. The definitions of urban and rural demographically, geographically, and economically vary from nation to nation (Tacoli 1998). In OECD countries, the process of defining rural and urban was introduced in two stages (Bollman 2005):

Demographically, the 'rural communities' or 'urban communities' classification is based on whether the population density was less than, greater than, or equal to 150 inhabitants per square kilometre. For most OECD countries, this threshold has not changed once it had been chosen.

Geographically, urban and rural regions are classified by the following index as 'predominantly rural regions' where 50% or more of their population lives in rural communities, while in 'intermediate regions' where 15–49% of their population lives in rural communities and 'predominantly urban regions' where less than 15% of their population living in rural communities.

However, this threshold in most Asian countries such as the Philippines was 500 persons per square kilometre. Normally in developing countries, rural means the countryside concerning country or village life and has a broader meaning than agrarian, which focuses on farming and agriculture economically. Whereas urban means cities and their lifestyle that is much different from a rural one, economically, it depends on industry and entertainments. In a word, rural areas are seen as an opposite to urban ones in developing countries (Antrop 2000; Wen et al. 2012).

Ecology is originally the branch of biology that attempts to explain the relationship between organisms and their habitats; it attempts to integrate the theory of all of science to investigate the structure and function of ecosystems (Brown 2009; Holdgate 1994; McDonnell and Pickett 1990). The primary objective of ecology is the conservation of biota and ecological processes under ecosystem functions. Conversely, ecology has been expanded to the fields of social sciences; particularly it has been integrated into bio-politics, agri-food, the field of nature-society interdependence (Goodman 1999; Vandermeer 1995). Essentially, when compared with 'urban', 'rural' is typically an ecosystem of social-ecological interdependence. Ecology is a more suitable vehicle to explain and advance human influenced landscapes (Wojtkowski 2004); integrating ecology into rural studies can bridge the gaps among farmers, agronomists, and ecologists.

Understanding 'rural' and 'ecology' contexts and various agricultural development in rural areas is valuable for understanding rural sustainability (Ellis and Biggs 2001); however, most of those themes or issues in rural studies during 1950–2010, such as organic farming, local food system, and rural agro-tourism, tended to be case studies. There has been little attempt to integrate them into a coherent conceptual framework to relate them to larger ecological, social, political, and ecological trends occurring at the national and global scales. Therefore, rural ecology tends to be defined as integrating ecology theory and system thinking methods to investigate the structure and function of rural ecosystems or agricultural landscapes in a rural context affected simultaneously by social, economic, cultural, and ecological factors.

Di Castri and Hadley (1986) summarized the meaning of the terms, multi-disciplinarity, pluri-disciplinarity, interdisciplinarity, or trans-disciplinarity that specified levels of coordination and interaction among disciplines related to ecological works. It also provides implications for using terms in agroecology (Dalgaard et al. 2003) and rural ecology. These are

- Multi-disciplinarity, several disciplines are involved but with no interaction.
- Pluri-disciplinarity, the uncoordinated interaction between disciplines.
- Interdisciplinarity, working across traditional disciplinary boundaries, is classified by:
- Unidirectional interdisciplinarity (interaction and coordination are imposed by a single discipline) and goal-oriented interdisciplinarity (interaction and coordination determined by the nature of an identified problem).
- Trans-disciplinarity exists where interaction involves not only the scientific and technological disciplines in stated goals but also where planners and administrators become involved in the processes.

Experience indicates that virtually all environmental issues are issues that transcend single discipline bounds. Rural ecology would be classified either as goal-oriented interdisciplinarity or as trans-discipline rather than just putting multi-disciplines together (Fig. 6.2). In other words, rural ecology can be developed as an interdisciplinary mode but with transdisciplinary analytical approaches. As

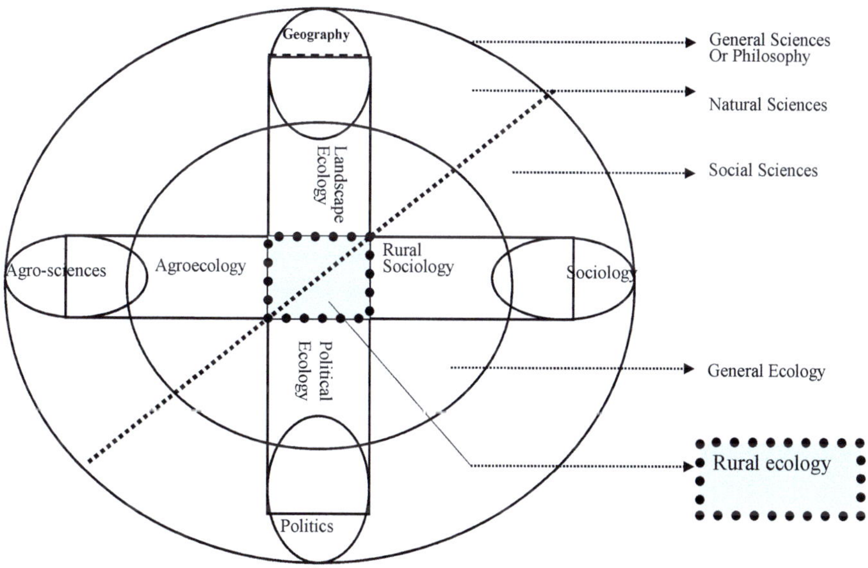

Fig. 6.2 Conceptual model of the trans- and interdisciplinarity of rural ecology

indicated in Fig. 6.2, in a modern science or discipline system, general science or philosophy is made up of social and natural science; modern ecology, dealing with the relationship between living organisms and their physical environment step into post natural dimensions and social dimensions (Holdgate 1994; Jarosz 2012), positions at the core of general science and bridges the gap between natural and social science. In general, rural ecology can be positioned as both a trans-discipline and an inter-discipline of many disciplines concerning rural development that interact with sociology, agricultural sciences (agronomy), politics, and geography and inside the scope of modern ecology, the most important four of those are political ecology, agroecology, landscape ecology, and rural sociology according to rural contextual society as society of an agriculture-orientated landscapes. Logically, this conceptual model also shows the trans- and interdisciplinarity of those four branches of ecology that intersect in rural ecology. Among which, rural sociology is a branch of sociology that deals with the relationship between the rural residents and social resource (Ferrelraj and Peixoto 1992; Beaulieu 2005). Currently, sustainability has been a central issue of rural sociology, and environmental factors have gradually been integrated in the context (Marsden et al. 2001; Ellis and Biggs 2001; Buttel 1996). Therefore, rural ecology, as a discipline to service rural sustainable development, needs to embrace rural sociology in the modern context of sustainability. Rural ecology will be a disciplinary framework for understanding ecosystem science contributions to the process of rural sociology.

Political ecology is the product of using ecology's ideologies to reconsider the basic elements of political organization and policy regarding sustainable resource management for future generations (Somma 1993; Turner 2004; Jarosz 2012). From urban to rural areas, most problems are related to politics and policy, the latter have been recognized in the management of the rural environment (Tricart and Kiewietde 1992; Vandermeer 1995). The nature of rural ecosystems has been largely dictated by agricultural policy to shape the modern agricultural industry (Green 1989; Landa et al. 1997;). The policy on land rights in most developing countries varies and poses dramatically different effects on land use pattern and rural livelihood (van Zanten et al. 2014); the lack of land tenure security and disappearing markets have caused considerable change in most rural areas of developing countries. For instance, in Mexico, many people blame development policy in rural areas for the country's environment problems, such as desertification, soil erosion, and landscape fragmentation (Landa et al. 1997). The triggers of 'Three Dimensional Rural Issues (TDRI)' in China (Wen 2001), has been considered to be the result of the long-term historical political institutional arrangement. Rural ecology aims to address such issues to gain understanding and to clear the bureaucratic barriers affecting natural and social properties in rural areas. For example, currently, land tenure in Chinese rural area can be an urgent issue of rural ecology related to political ecology.

Landscape ecology, initiated by Carl Troll in 1939 just 4 years after the ecosystem concept was proposed by A.G. Tansley in 1935 (Haber 2004; Turner 2005), is an inter-discipline of ecology and geography that focuses on the reciprocal interactions between spatial pattern and ecological process (Turner 2004). Because landscape transformation is a core trait from geographic changes to economic results both in urban and rural context (Moss 2000), its precepts and principles became increasingly accepted in other disciplines accommodating both natural and social landscape dimension and expanding to culture landscape. Naveh (1995) proposed a holistic concept of landscape ecology as a transdisciplinary science to embrace natural, social, and cultural landscape. Haber (2004) further defined landscape as the science of environmental interactions which act as bridges between humanity and social philosophy. Essentially, those landscapes are driven naturally via solar power and socially via information power (Naveh 1998). Landscape ecology approaches aim to address the studies of different scales, organisms, and processes (Zimmerer 2004, 2006; Mathewson 1999); thus, social and cultural landscape analysis has been an indispensable part in analysing the framework of landscape appraisal with natural landscape for rural areas (Field et al. 2003). Generally, rural areas are agriculture-orientated landscapes, and many rural-based problems stem from the agricultural landscape shifts, such as soil erosion, cropping system regulation, deforestation, resource degradation, food insecurity, and rural poverty. Therefore, landscape agroecology, a new branch of agroecology, was initiated as the ecology of a productive countryside (Wojtkowski 2004). Conversely, in most developing countries, rural-based problems have both natural and social orientations and each becomes the cause of another. In this respect, landscape analysis should lie at the core of rural ecology that integrates natural, social, and cultural landscape analysis. Social landscape analysis can be defined as the study of spatial distribution of the interrelated social variables in each biophysical setting to help researchers understand the social causes of natural problem and vice versa. Moreover, because land use shift has the largest effect on biodiversity, this determines, to a large extent, the structure and function of rural ecosystems (Chapin III and Walker 1997). Landscape analysis is the basis of the analytical framework of rural ecology.

Agroecology has been defined as the unification of ecological principles to address biological and systematic problems of agriculture for more than 80 years (Dalgaard et al. 2003). It has become a research and developmental paradigm for world agriculture (Altieri 1989), and since the concept and philosophy of Sustainable Development were proposed in 1987, agroecology has been developed to be the central discipline of sustainable agriculture (Thomas and Kevan 1993) and food systems (Francis 2003). In developing countries, agroecology can be applied to enhance the productivity of farming systems to alleviate rural poverty and hunger (Altieri 1999). Agroecological practice will need to act as a component of the new condition of rural sustainability instead of production agriculture (Marsden et al. 2001); thus, we can integrate agroecological approaches creatively to build upon social constructivism for rural development (Marsden et al. 2001). However, in most developing countries, agroecology has not had sufficient disciplinary development and course teaching, despite some significant efforts such as those in China to restructure agroecology subjects as a central discipline of future agriculture shifts to sustainable, knowledge-based agriculture (Wang et al. 2019). Agroecological perspective can give a clearer direction in analysing human-nature co-evolution and interacting rural development. With that in mind, it is reasonable to develop rural ecology under the agroecological discipline for future rural development in developing countries.

In summary, almost all processes in rural areas are related both socially and ecologically. Rural ecology is developed as a central discipline to address such socio-ecological processes rather than a single discipline like rural sociology or agricultural economics. Issues and themes in rural development have greatly changed from unique social relationships to multidimensional ones related to the sustainable livelihoods in rural area (Ellis and Biggs 2001). Ecology, as bridge discipline of natural and social sciences, is now ready to provide modernization to enhance rural development. Agroecology, as an applied branch of ecology, will be integrated with rural ecology to address the rural landscape and relationship; therefore, it is time for rural ecology to be active in the twenty-first century, so that the agro-landscape-social-ecological-cultural factors can be integrated into future studies of rural sustainable development and rural livelihood.

3.2 *Scales of Rural Ecology*

Scale means the spatial and temporal quantitative or analytical dimensions used to measure and study a scientific phenomenon. Dalgaard et al. (2003) stated that the scale and interdisciplinarity of agroecology as a separate discipline must be addressed; to do that, they characterized both 'soft' and 'hard' agroecology. 'Soft' agroecology means 'system thinking' aspects that function or work at a large scale, while 'hard" agroecology encompasses the technical or engineering dimensions integrated from ecology, agronomy, and economy that would function in the small scale. Rural ecology can also consist of such 'soft' and 'hard' thinking when considering its similarities with agroecology. As stated above, rural ecology needs to be addressed in the natural, social, and cultural landscapes. Natural landscape would be considered a 'hard' aspect using ecological approaches for its transformation, while social relation would be a 'soft' aspect exploring system thinking to make correct actions of the former (Packham and Sriskandarajah 2005; Marsden et al. 2001; Ellis and Biggs 2001).

In doing so, we draw out both soft and hard scales of rural ecology in coordination with urban ecology (Fig. 6.3). Generally, ecology is developed as human ecology or ethnoecology describing how humans interact with their habitat (Altieri 2002; Tjallingii 2000; Somma 1993; Hobbs and Morton 1999; Carolan 2005). To some extent, human ecology is 'soft' ecology addressing the philosophy of human views of nature (Becker and Ostrom 1995), while ethnoecology is 'hard' ecology to create a series of technical and systematic approaches to realize relationships with nature. Geographically, as shown in Fig. 6.3, ethnoecology can be divided by urban ecology and rural ecology; the discussion of the former is not in the scope of this paper. Rather, attention is concentrated on the scales of rural ecology, which is

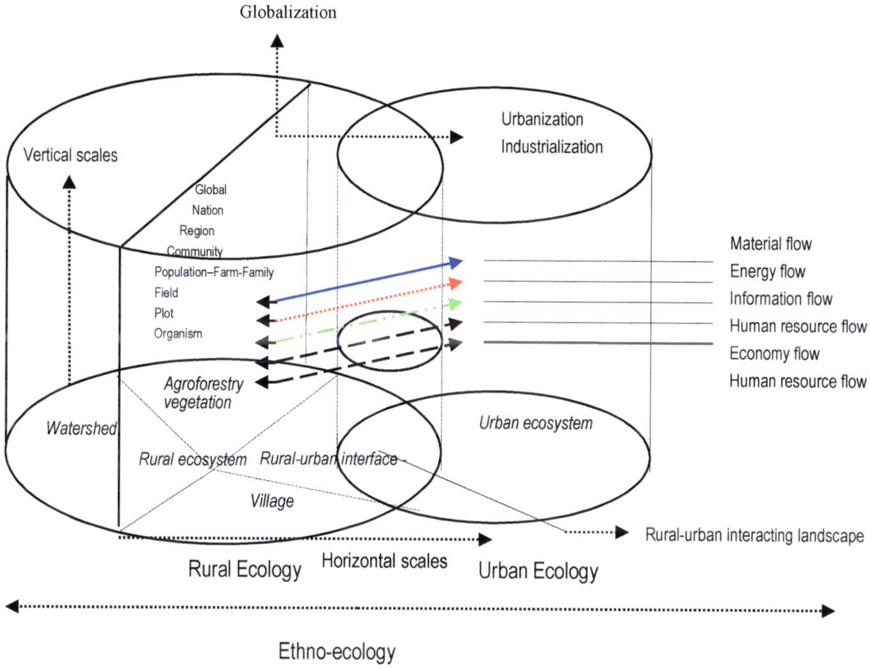

Fig. 6.3 Scales of rural ecology related to urban ecology in the spectrum of general ethnoecology geographically

complementary to urban ecology that has already been a separate discipline more than 30 years (Mcintyre 2000, r99; Rebele 1994).

As stated above, rural landscape analysis will be the core element of rural ecology. Rural landscape can be interpreted at structural and functional aspects (Claval 2005); geographically, it includes horizontal scale and vertical scales.

3.2.1 Horizontal Scales

According to Claval (2005), traditional rural landscapes can be interpreted via three perspectives: (i) agricultural fields such as meadows, pastures, and woods; (ii) inhabited societies; and (iii) natural surfaces. Accordingly, we would like to divide rural landscape geographically by:

Village landscape, being a rural resident centralized landscape, placing households in rural studies with emphasis on social and cultural dimensions (Zimmerer 2004).

Agroforestry landscape, organized for agricultural production lying typically in between natural and social aspects.

Watershed landscape, tending to be natural dimension but more influenced by agricultural practice (Claval 2005).

Rural-urban interacting landscape (perhaps the most relevant to this discussion) being the interface between rural and urban landscape (Battisti and Gippoliti 2004; McDonnell and Pickett 1990; Tacoli 1998), to some degree, the structure and function of rural ecosystem are determined by material, energy, economic, and information (including intelligence and human resource in social dimension) flows exchanged at this interface. The increasingly urbanizing process may produce a variety of unprecedented effects for rural development in developing countries (McDonnell and Pickett 1990).

The focus of rural ecology should be on the relationship of the village landscape with the interaction between the rural and urban landscapes (or ecosystems). There are sophisticated structural and functional determinants that should be the focus of rural ecology. They influence each other both in structure and function by flows of material, energy, and information. The most important consideration is that Human Resources need to be addressed in Rural ecology. Therefore, we now define rural ecology as a science of relationship between the living systems and their environment in rural landscapes. The core of this is the village landscape and its interaction with urban ecosystems by material, energy, information, economy, and human resource flows.

3.2.2 Vertical Scales

The vertical scale of rural ecology is determined by hierarchical scales of certain ecosystems (Haber 2004). As research objects of rural ecology, rural ecosystem rural ecosystems are hierarchically classified as Global, National, Regional, Community, Population (farm or family household), and Organism (individual). Currently, global is a very vital issue with economy and information globalization having great effects on rural landscape shifts in developing countries.

3.3 Analytical Framework of Rural Ecology for Developing Countries: A Case of Rural China

China is one of the most politically, economically, and environmentally influential countries in the world. China is now facing a heavy environmental crisis of enormous proportions after several decades of fast economic growth (Liu and Diamond 2005). Its urban environmental problems are more like that of industrial countries that suffered from urbanization and industrialization in the past, whereas the rural environmental problems, the so-called 'Three-dimensional Rural Issues of "Agriculture", "Farmer" and "Countryside" are much more specific and urgent. Such issues have been suffering from "Dualisms', i.e. the Nature vs Society Dualism and Urban vs Rural Dualism (or Divide) (Tickell 2001, 2004; Knight and Song 1999). Therefore, rural ecology is designed to offer effective analytical approaches to address such multisectoral problems.

3.3.1 Problems and Challenges in Rural China

China is the world's second most populous country and the fourth largest in area (Liu and Diamond 2005). Nearly 70% of the 1.3 billion people still live in rural areas due to the long established Rural-Urban Divide policy. Rural residents live in villages and rural townships in the vast and relatively underdeveloped interior; this makes China essentially a nation of villages (Pant 1999). In this regard, rural ecology is a particularly necessary discipline for Chinese academia (Zhou et al. 1999, r4). It is said that China has achieved food supply for 22% of world population using 7% of the world's land, but this has been achieved at considerable social and environmental costs to rural China. Large areas have been deforested for food production; vast water projects and dams have been used for agricultural irrigation; the land they rely on is not only small in area per capita (less than 0.07 hm^2) but even much degraded. Much of the agricultural land is becoming more and more unsuitable for farming due to the increasingly intensive farming and chemical input for pursuing the maximum outcome annually. The situation is getting worse under attack of the world free trade trend after China's accession to the WTO (Anbumozhi and Radhakrishnan 2005). Nutrient inputs have lost balance within the agroecosystems to get more from the rural ecosystem. Presently, the negative effects of industrialization, urbanization, and capitalist economy are not only in the urban area but also are expanding to the rural areas (Han 1989; Tickell 2004); land and rivers are contaminated, water is deteriorated and depleted, and food is increasingly unsafe.

The main challenges in rural areas of China relate to the 'complex interactions of economics, ecology and sociology, (Lin et al. 2003; Tickell 2001; Liu and Diamond 2005; Zhen et al. 2005; Huang et al. 2005). Such problems are triggering the 'three-dimensional rural issues' (Wen 2001) concerning agriculture, farmer, and rural environment simultaneously. Disciplinarily, it obviously involves both social and natural (ecological) processes and therefore needs the participation of inter- or transdisciplinary cooperation between the natural and social sciences. Sustainable development remains a huge challenge in Chinese rural areas. The food production system has directly led to rural landscape change (Palang 2005), which in turn moulds Chinese economic and social patterns. The application of rural ecology can redefine these relationships, nature and society, landscape and ecology, holism and dichotomy, theory and practice, to move it towards sustainable development and make a difference in China. Rural ecology will focus on various spatial scales (global, regional, local, etc. see Fig. 6.3) and promote a mechanism whereby a holistic, integrated, and long-term approach to the rural sustainable development. In doing so, economic, social, and environmental factors will be integrated in an analytical framework of rural ecology operating at different spatial and temporal scales, demonstrating how environmental issues might be integrated with social, economic, and geographic results.

3.3.2 Analytical Framework of Rural Ecology for Rural China

Agriculture is one of the sectors that are prone to failure due to mistakes in marketing and policy (Long et al. 2003). It is very closely associated with environmental quality by its impacts on land use, water resource, and biodiversity influencing crop production patterns, farming practices, and chemical inputs.

Essentially, the basic understanding of rural sustainability is to integrate the essence of ecological, economic, and social science theory as an analytical framework to throw insight into the complexity and wholeness of rural ecosystems. This is the main purpose for proposing a newly constructed rural ecology; below is presented a rural ecology framework and approach to bridge the gap between rural society and environment in China.

3.3.3 Sustainability in Rural China: Landmark for Rural Ecology

The concept of sustainability has been interpreted in many dimensions and contexts (Rocheleau 1991). Generally, sustainable development means the kind of development that meets the needs of the present without harming the next generation's ability to meet their own needs (WCED 1987; Goodland and Daly 1996), whereas practically, sustainable development refers to the goal of fostering adaptive capacities and creating opportunities that create, test, and maintain adaptability (Holling 2001). Relative to landscape ecology, the concept and connotation of rural sustainability remains vague; in the view of rural ecology, rural sustainability depends on the interactions among internal and external factors, including social, political, economic, and ecological dimensions both inside and outside the rural ecosystem (Akpinar et al. 2004). This approach provides a landmark approach to rural ecology for China.

Regarding the reality of the rural China, rural sustainability needs to be changed from a deceptive buzzword to a critical framework. Appropriately balanced attention to economic, ecological, and equity factors or issues should be taken into account in the rural development decision process, while giving due consideration to the future (even more distant future) states.

According to Goodland (1995) and Goodland and Daly (1996), sustainability consists of three contextual layers: environmental sustainability, economic sustainability, and social sustainability. These are consistent with the three-dimensional problems analysed above for rural China and refer to environmental soundness (the preservation and improvement of the natural environment), economic viability(the maintenance of yields and productivity of crops and livestock), and social acceptability (self-reliance equality and improved quality of life), respectively (Zhen et al. 2005). Basically, environmental sustainability is fundamental to economic sustainability and the combination of these becomes a prerequisite of social sustainability. Analytical frameworks in the context of rural China within scope of rural ecology are summarized in Table 6.1, both as a branch of ecology and as a distinct discipline.

Table 6.1 Environmental, economic, and social sustainability in context of rural China in rural ecology and their analytical frameworks

Components of sustainability	Environmental sustainability	Economic sustainability	Social sustainability
Structural characteristics	Natural relations	Human-made-nature relations	Human-nature relations
Ecological process	Natural capital (material and energy) flow	Human made (merchant capital flow)	Human and social capital flow
Ecosystem functions	Maintenance of natural capitals	Maintenance of human-made (economic) capitals	Maintenance of human and social capitals

Rural ecology will focus on:

Rural ecosystem structure, function, and their dynamic process: Interacting regimes between landscape and living beings ecologically (physically, chemically, physiologically, climatologically, and genetically) and sociologically (culturally, economically, and politically).

Rural landscape structure and dynamic: Agricultural landscapes are one dimension of agricultural systems which express visually many of the results of the interactions of a multiplicity of factors, societal, economic, historic, and environmental (Moss 2000). Land use changes frequently happen due to rural agricultural exploitation and alter the abundance of biodiversity, especially those that influence water and nutrient dynamics, trophic interactions that determine the structure and functioning of rural ecosystems (Chapin III and Walker 1997).

Rural-urban interacting regimes: Structure and function of whole rural ecosystems are largely determined by processes occurring at this interaction point; the ecological and social effects on rural development of globalization, urbanization, and industrialization will be highlights of future studies of rural ecology.

3.3.4 Analytical Indicators of Sustainability for Rural China

The concept of rural ecosystems is much wider than agroecosystem (Flora 2001), however, when considering the reality of rural China, agriculture has been a centre of rural livelihood and the sustainability indicator of certain agroecosystem can be used for appreciating most rural ecosystems. These are:

Productivity, recognizing the productive ability of rural ecosystems, not only in terms of efficiency measures of economically valuable production of food and fibre but also in terms of the ability to provide service functions (e.g. soil maintenance, water conservation aesthetic landscape) of rural ecosystem. It can be a quantitative measure and is a useful indicator to compare functions between different types of rural ecosystems.

Stability refers to the maintaining ability of productivity when a rural ecosystem is affected by normal fluctuation of ambient factors (biological, physical, marketing, etc.). Therefore, stability can be designed by coefficient of variance (CV) of productivity statistically.

Sustainability, specific in its narrow sense, refers to the maintaining ability of productivity when a rural ecosystem is interfered by abnormal fluctuation of anthropogenic factors (e.g. soil salinization, soil erosion, flood, financial crisis, etc.), such fluctuation can be abrupt by the accumulated effects of human unfavourable activities that can make a decrease or even loss of the productivity and service of rural ecosystem, thus to be able to reflect slight difference of rural sustainability in broad sense.

Homogeneity is explored to reflect the concerted degree between functional sectors within rural ecosystems which generally include planting, forestry, pasture, fishery and sideline (off-farm) in traditional rural China. Ecologically, homogeneity is very fundamental to differentiate and appreciate the functional flow of rural ecosystems.

Equity means the evenness of resource and profit distributed among the rural participants, quantitatively, it can be formulated by the reciprocal of coefficient of variance (CV) of resource allocation individually; socioeconomic equity is a fundamental element of rural sustainability (Daily and Ehrlich 1996) and obviously equity is a critical indicator for sustainability in rural China.

References

Addinsall, Cherise, Kevin Glencross, Pascal Scherrer, Betty Weiler, and Doland Nichols. 2015. Agroecology and sustainable rural livelihoods: a conceptual framework to guide development projects in the Pacific Islands. *Agroecology and Sustainable Food Systems* 39 (6): 691–723.

Akpinar, N., I. Talay, C. K. Ceylan, and S. Gunduz. 2004. Rural women and agrotourism in the context of sustainable rural development: a case study from Turkey. *Kluwer Journal* 6:473–486.

Altieri, M. A. 1989. Agroecology: a new research and development paradigm for world agriculture. *Agriculture, Ecosystems and Environment* 27:37–46.

Altieri, M. A. 1993. Ethnoscience and biodiversity: key elements in the design and of sustainable pest management systems for small farmers in developing countries. *Agriculture, Ecosystems and Environment* 46:257–272.

Altieri, M. A. 1995. Sustainable rural development. In *Agroecology: the science of sustainable agriculture*, ed. M. A. Altieri, 64–68. Boulder, CO: Westview Press.

Altieri, M. A. 1998. An agroecological perspective to guide graduate educational programs in agricultural economics and rural development in Latin America of the XXI century. *Ecological Economics* 27:227–236.

Altieri, M. A. 1999. Applying agroecology to enhance the productivity of peasant farming systems in Latin America. *Environment, Development and Sustainability* 1:197–217.

Altieri, M. A. 2002. Agroecology: the science of natural resource management for poor farmers in marginal environments. *Agriculture, Ecosystems and Environment* 93 (1–3): 1–24.

Amekawa, Yuichiro. 2011. Agroecology and sustainable livelihoods: towards an integrated approach to rural development. *Journal of Sustainable Agriculture* 35 (2): 118–162.

Anbumozhi, V., and J. Radhakrishnan. 2005. Effects of agricultural free trade on environmental quality: exploring the linkages and research perspectives. *Outlook Agriculture* 34:91–96.

Antrop, M. 2000. Changing patterns in the urbanized countryside of western Europe. *Landscape Ecology* 15:257–270.

Barrios, Edmundo, Barbara Gemmill-Herren, Abram Bicksler, Emma Siliprandi, Ronnie Brathwaite, Soren Moller, Caterina Batello, and Pablo Tittonell. 2020. The 10 elements of agroecology: enabling transitions towards sustainable agriculture and food systems through visual narratives. *Ecosystems and People* 16 (1): 230–247. https://doi.org/10.1080/26395916.2020.1808705.

Battisti, C., and P. Gippoliti. 2004. Conservation in the urban-countryside interface: a cautionary note from Italy. *Conservation Biology* 18:581–583.

Beaulieu, L. J. 2005. Breaking walls, building bridges: expanding the presence and relevance of rural sociology. *Rural Sociology* 70:1–27.

Becker, C. D., and E. Ostrom. 1995. Human ecology and resource sustainability: the importance of institutional diversity. *Annual Review of Ecology and Systematics* 26:113–133.

Bensin, B.M. 1928. *Agroecological characteristics description and classification of the local corn varieties chorotypes*. Book, (Publisher unknown so far. Cited from: Wezel A, Bellon S, Doré T, Francis C, Vallod D, David C (2009) Agroecology as a science, a movement and a practice. A review. Agron Sustain Dev. 29(4):503–515.)

Bollman, R. D. 2005. Rural Canada in the OECD context. *Journal of Agricultural Safety and Health* 11 (2): 151–157.

Brown, Joel R. 2009. Climate change and ecology in rural lands. In *The planner's guide to natural resource conservation*, ed. A. X. Esparza and G. McPherson, 39–56. Cham: Springer. https://doi.org/10.1007/978-0-387-98167-3_3.

Brush, S. B. 1989. Rethinking crop genetic resource conservation. *Conservation Biology* 3:19–29.

Buttel, F. H. 1996. Environmental and resource sociology: theoretical issues and opportunities for synthesis. *Rural Sociology* 61:56–76.

Carolan, M. S. 2005. Society, biology, and ecology: bringing nature back into sociology's disciplinary narrative through critical realism. *Organization and Environment* 18:393–421.

Chapin, F. S., III, and B. H. Walker. 1997. Biotic control over the functioning of ecosystems. *Science* 277. https://doi.org/10.1126/science.277.5325.500.

Claval, P. 2005. Reading the rural landscapes. *Landscape and Urban Planning* 70 (1/2): 9–19.

Daily, G. C., and P. R. Ehrlich. 1996. Socioeconomic equity, sustainability, and earth's carrying capacity. *Ecological Applications* 6 (4): 991–1001.

Dalgaard, T., N. J. Hutchings, and J. R. Porter. 2003. Agroecology, scaling and interdisciplinarity. *Agriculture, Ecosystems and Environment* 100:39–51.

De Schutter, Olivier. 2011. *Agroecology and right to food.* Report presented at the 16th Session of the United Nations Human Rights Council [R/OL]. Human Rights Council of the United Nations. http://www.srfood.org/images/stories/pdf/officialreports/20110308_a-hrc-16-49_agroecology_en.pdf. Accessed June 17, 2024 in Canada.

Di Castri, F., and M. Hadley. 1986. Enhancing the credibility of ecology: is interdisciplinary research for land use planning useful? *GeoJournal* 13:299–325.

Ellis, F., and S. Biggs. 2001. Evolving themes in rural development 1950s-2000s. *Development and Policy Review* 19 (4): 437–448.

FAO. 2017. *The state of food and agriculture 2017: leveraging food systems for inclusive rural transformation*. Rome.

FAO. 2024. *The state of food security and nutrition in the world 2024*. https://www.fao.org/publications/fao-flagship-publications/the-state-of-food-security-and-nutrition-in-the-world/en. Accessed April 14, 2025.

Ferrelraj, M. C., and J. Peixoto. 1992. Rural sociology and rural development in Portugal history, recent trends and prospects. *Irish Journal of Sociology* 2:122–141.

Field, D. R., P. R. Voss, T. K. Kuczenski, R. B. Hammer, and V. C. Radeloff. 2003. Reaffirming social landscape analysis in landscape ecology: a conceptual framework. *Society and Natural Resources* 16:349–361.

Flora, C., ed. 2001. *Interactions between agroecosystems and rural communities*. Boca Raton, FL: CRC Press.

Francis, C., G. Lieblein, S. Gliessman, T. Breland, N. Creamer, R. Harwood, L. Salomonsson, J. Helenius, D. Rickerl, R. Salvador, M. Wiedenhoeft, S. Simmons, P. Allen, M. Altieri,

C. Flora, and R. Poincelot. 2003. Agroecology: the ecology of food systems. *Journal of Sustainable Agriculture* 22:99–118.

Goodland, R. 1995. The concept of environmental sustainability. *Annual Review of Ecology and Systematics* 26:1–24.

Goodland, R., and H. Daly. 1996. Environmental sustainability: universal and non-negotiable. *Ecological Applications* 6:1002–1017.

Goodman, D. 1999. Agro-food studies in the 'age of ecology': nature, corporeality, bio-politics. *Sociologia Ruralis* 39:17–38.

Green, B. H. 1989. Agricultural impacts on the rural environment. *Journal of Applied Ecology* 26:793–802.

Gumbo, D. J., M. Dumas-Johansen, G. Muir, F. Boerstler, and Z. Xia. 2018. *Sustainable management of Miombo woodlands—food security, nutrition and wood energy*. Rome: Food and Agriculture Organization of the United Nations. http://www.fao.org/3/i8852en/I8852EN.pdf. Accessed July 25, 2024.

Haber, W. E. 2004. Landscape ecology as a bridge from ecosystems to human ecology. *Ecological Research* 19:99–106.

Haila, Y. E. 1989. Ecology: finding evolution finding ecology. *Biology and Philosophy* 4:235–244.

Han, C. R. 1989. Recent changes in the rural environment in China. *Journal of Applied Ecology* 26:803–812.

Hayami, Y. 2001. Ecology, history, and development: a perspective from rural Southeast Asia. *World Bank Research Observer* 16:169–198.

Hobbs, R. J., and S. R. Morton. 1999. Moving from descriptive to predictive ecology. *Agroforestry Systems* 45:43–55.

Holdgate, M. W. 1994. Ecology, development and global policy. *Journal of Applied Ecology* 31:201–211.

Holling, C. S. 2001. Understanding the complexity of economic, ecological, and social systems. *Ecosystems* 4:390–405.

Hooker, N. H. 1999. Food safety regulation and trade in food products. *Food Policy* 24:653–668.

Huang, Q. Q., D. Dawe, S. Rozelle, J. K. Huang, and J. X. Wan. 2005. Irrigation, poverty and inequality in rural China. *The Australian Journal of Agricultural and Resource Economics* 49:159–175.

Jackson, Dana L., and Laura L. Jackson Johnson. 2002. *The farm as natural habitat: reconnecting food systems with ecosystems*, 207. Washington, DC: Island Press.

Jarosz, Lucy. 2012. Growing inequality: agricultural revolutions and the political ecology of rural development. *International Journal of Agricultural Sustainability* 10 (2): 192–199.

Julia-Serres, Bailey, Jane E. Parker, Elizabeth A. Ainsworth, Giles E. D. Oldroyd, and Julian I. Schroeder. 2019. Genetic strategies for improving crop yields. *Nature* 575:109–118.

Jung, Da-Jung, and Sung-Sang Yoo. 2023. Reacting to rural problems or creating cognitive space? Knowledge production and the diffusion of the agroecology movement in Chiang Mai, Thailand. *Agroecology and Sustainable Food Systems* 47 (8): 1186–1206. https://doi.org/10.1080/21683565.2023.2230928.

Knight, J., and L. Song. 1999. *The rural-urban divide: economic disparities and interactions in China*. New York: Oxford University Press.

Landa, R., J. Meave, and J. Carabias. 1997. Environmental deterioration in rural Mexico: an examination of the concept. *Ecological Applications* 7:316–329.

Leopold, Aldo. 1949. *A Sand County almanac*. New York: Oxford University Press.

Lin, Yifu, Guo Guodong, Li Li, et al. 2003. *Long-term growth and prospects of China economy*. Peking University working paper.

Liu, J. G., and J. Diamond. 2005. China's environment in a globalizing world. *Nature* 435:B1179–B1186.

Long, C. L., H. Li, Z. Q. Ouyang, X. Y. Yang, Q. Li, and B. Trangmar. 2003. Strategies for agrobiodiversity conservation and promotion: a case from Yunnan, China. *Biodiversity and Conservation* 12:1145–1156.

Marsden, T., J. Banks, H. Renting, D. P. Van, and D. Jan. 2001. The road towards sustainable rural development: issues of theory, policy and research practice. *Journal of Environmental Policy and Planning* 3:75–83.

Mathewson, K. 1999. Cultural landscape and ecology II: regions, retrospects, revivals. *Progress in Human Geography* 23:267–281.
McDonnell, M. J., and S. T. A. Pickett. 1990. Ecosystem structure and function along urban-rural gradients: an unexploited opportunity for ecology. *Ecology* 71:1232–1237.
Mcintyre, N. E. 2000. Urban ecology as an interdisciplinary field: differences in the use of "urban" between the social and natural sciences. *Urban Ecosystem* 4:5–24.
Méndez, V. Ernesto, Christopher M. Bacon, and Roseann Cohen. 2013. Agroecology as a transdisciplinary, participatory, and action-oriented approach. *Agroecology and Sustainable Food Systems* 37 (1): 3–18.
Midmore, P., and J. Whittaker. 2000. Economics for sustainable rural systems. *Ecological Economics* 35:173–189.
Moss, M. R. 2000. Interdisciplinarity, landscape ecology and the transformation of agricultural landscapes. *Landscape Ecology* 15:303–311.
Naveh, Z. 1995. Interactions of landscapes and cultures. *Landscape and Urban Planning* 32:43–54.
Naveh, Z. 1998. Ecological and cultural landscape restoration and the cultural evolution towards a post-industrial symbiosis between human society and nature. *Restoration Ecology* 6 (2): 135–143.
Packham, R., and N. Sriskandarajah. 2005. Systemic action research for postgraduate education in agriculture and rural development. *Systems Research and Behavioral Science* 22:119–130.
Palang, H., and Helmfrid. 2005. Rural landscapes: past processes and future strategies. *Landscape and Urban Planning* 70:3–8.
Pant, D. R. 1999. On China's villages, rural industries and technology development: a review essay with special reference to technology development in rural industries. *Journal of Political Ecology-Case Studies in History and Society* 6:193.
Rebele, F. 1994. Urban ecology and special features of urban ecosystems. *Global Ecology and Biogeography Letters* 4:173–187.
Rocheleau, D. E. E. 1991. Gender, ecology, and the science of survival: stories and lessons from Kenya. *Agriculture and Human Values* 8:156–165.
Rosset, P. M., and M. E. Martínez-Torres. 2012. Rural social movements and agroecology: context, theory, and process. *Ecology and Society* 17 (3): 17.
Rover, Oscar José, Bernardo Corrado de Gennaro, and Luigi Roselli. 2017. Social innovation and sustainable rural development: the case of a Brazilian agroecology network. *Sustainability* 9 (3): https://doi.org/10.3390/su9010003.
Somma, M. 1993. Theory building in political ecology. *Social Science Information* 32:371–385.
Tacoli, C. 1998. Rural-urban interactions: a guide to the literature. *Environment and Urbanization* 10:147–166.
Thomas, V. G., and P. G. Kevan. 1993. Basic principles of agroecology and sustainable agriculture. *Journal of Agricultural and Environmental Ethics* 6:1–19.
Tickell, C. 2001. China on the losing side: Mao's drive to conquer nature had shocking consequence of his country. *Nature* 412:273–274.
Tickell, C. 2004. The decline of China's environment: the spread of agriculture led to deforestation and growth of towns. *Nature* 430:505–506.
Tilzey, Mark. 2024. Food democracy as radical political agroecology: securing autonomy (alterity) by subverting the state-capital nexus. *Frontiers in Sustainable Food Systems* 8:1044999. https://doi.org/10.3389/fsufs.2024.1044999.
Tjallingii, S. P. 2000. Ecology on the edge: landscape and ecology between town and country. *Landscape and Urban Planning* 48:103–119.
Tornaghi, Chiara, and Michiel Dehaene. 2020. The prefigurative power of urban political agroecology: rethinking the urbanisms of agroecological transitions for food system transformation. *Agroecology and Sustainable Food Systems* 44 (5): 594–610. https://doi.org/10.1080/21683565.2019.1680593.
Tricart, J., and J. Kiewietde. 1992. *Ecogeography and rural management: a contribution to the International Geosphere-Biosphere Programme*. Harlow: Longman Scientific & Technical, Copublished with John Wiley & Son, Inc.

Turner, M. D. 2004. Political ecology and the moral dimensions of "resource conflicts": the case of farmer-herder conflicts in the Sahel. *Political Geography* 23:863–889.

Turner, Monica G. 2005. Landscape ecology: what is the state of the science? *Annual Review of Ecology, Evolution, and Systematics* 36:319–344.

United Nations, Department of Economic and Social Affairs, Population Division. 2024. *The 2024 revision of world population prospects*. https://population.un.org/wpp/. Accessed April 11, 2025 (also see https://data.worldbank.org.cn/indicator/SP.RUR.TOTL).

van Zanten, Boris T., Peter H. Verburg, Maria Espinosa, Sergio Gomezy-Paloma, Giuliano Galimberti, Jochen Kantelhardt, Martin Kapfer, Marianne Lefebvre, Rosa Manrique, Annette Piorr, Meri Raggi, Lena Schaller, Stefano Targetti, Ingo Zasada, and Davide Viaggi. 2014. European agricultural landscapes, common agricultural policy and ecosystem services: a review. *Agronomy for Sustainable Development* 34:309–325.

Vandermeer, J. 1995. The ecological basis of alternative agriculture. *Annual Review of Ecology and Systematics* 26:201–224.

Vasishth, Ashwani. 2008. A scale-hierarchic ecosystem approach to integrative ecological planning. *Progress in Planning* 70:99–132.

Wang, S. L., and C. D. Caldwell. 2016. Ecology: a core science in ecological civilization era—exemplified with the roles of rural ecology in guiding the beautiful village construction (in Chinese). *Journal of Fujian Agriculture and Forestry University* 19 (3): 14–18.

Wang, S. L., C. D. Caldwell, S. L. Kilyanek, M. Sean, and Smukler. 2019. Using agroecology to stimulate the greening of agriculture in China: a reflection on 15 years of teaching and curriculum development. *International Journal of Agricultural Sustainability* 17 (4): 298–311.

Wang, S. L., S. X. Shi, R. Y. Wu, and Y. W. Dai. 2021. Rural ecology: an emerging discipline for the rural sustainable development (in Chinese). *Chinese Journal of Eco-Agriculture* 29 (12): 2116–2125.

Wen, Tiejun. 2001. Centenary reflections on the 'three dimensional problem' of rural China. *Internatioanl-Asia Cultural Studies* 2:287–295.

Wen, Tiejun, Kinchi Lau, Cunwang Cheng, Huili He, and Jiansheng Qiu. 2012. Ecological civilization, indigenous culture, and rural reconstruction in China. *Monthly Review* 63 (9): 29–35.

Wezel, Alexander, S. Bellon, T. Doré, C. Francis, D. Vallod, and C. David. 2009. Agroecology as a science, a movement and a practice. A review. *Agronomy for Sustainable Development* 29 (4): 503–515.

Wojtkowski, P.A. 2004. *Landscape agroecology*. Food Products Press, an inprint of The Haworth Press Binghamton, NY.

World Bank Group. 2024. *Poverty, prosperity, and planet report 2024*. https://www.worldbank.org/en/publication/poverty-prosperity-and-planet. Accessed April 11, 2025.

World Commission on Environment and Development. 1987. *Our common future*. Oxford: Oxford University Press.

Zhen, L., G. B. Thapa, and G. D. Xie. 2005. Agricultural sustainability in the food bowl of the North China plain. *Outlook Agriculture* 34:149–158.

Zhou, Daowei, Lianxi Sheng, Zhengfang Wu, and Wei Gao. 1999. On village ecology (in Chinese). *Chinese Journal of Applied Ecology* 10 (3): 369–372.

Zimmerer, K. S. 2004. Cultural ecology: placing house hold sin human-environment studies—the cases of tropical forest transitions and agrobiodiversity change. *Progress in Human Geography* 28:795–806.

Zimmerer, K. S. 2006. Cultural ecology: at the interface with political ecology—the new geographies of environmental conservation and globalization. *Progress in Human Geography* 30:63–78.

Chapter 7
New Challenges for Education in Agroecology

Abstract Based upon the experience of our international agroecological teaching and rural community engagement over two decades, this chapter delineates the nature of agroecology that can shed new light on higher education transformation in the world towards a sustainable global agri-food system.

Keywords Education in agroecology · Agricultural sustainability · Agriculture associated universities and colleges (AAUCs) · Competence education · The *agroecologists*

Learning Objectives
After reading this chapter, students should be able to:

1. Describe the essential elements of education in university and colleges to support future agricultural sustainability.
2. Discuss the challenges for agroecology education.
3. Describe the potential strategies of future education of agroecology in agriculture-associated universities and colleges (AAUCs).

****

> There is no wealth but life.
> —John Ruskin

> Education is not the learning of facts, but the training of the mind to think.
> —Albert Einstein

> Poor training of agricultural extension staff has been identified as part of the problem of the relative ineffectiveness of much of extension in the field.
> —Rogers A. (1996)

Every question and problem in our world can be linked back to education, both formal and informal, meaning that proper education can lead to the answer to each

S. Wang, C. D. Caldwell, *Advanced Agroecology*,
https://doi.org/10.1007/978-3-032-08689-1_7

question. Similarly, our current agri-food problems also relate to education, because every stakeholder in the agri-food system is educated in some manner, whether informally through experience and indigenous knowledge or via the myriads of available schools and colleges. Because in the last 100 years, indigenous knowledge has been replaced by top-down formal education in agriculture, the ideology and approaches to agricultural and food systems have become nested in our agricultural colleges and universities. These approaches have been critical in shaping the rural landscape and our effects on agroecosystems. Basically, in agriculture and food-associated colleges and universities, the courses and curriculum are designed and taught in discrete discipline boundaries, much like the five blindmen guessing the structure of an elephant, resulting in huge information loss of the whole agroecosystem. To address this problem, education in agroecology, which is by its nature an interdisciplinary combination of agronomy, economics, and ecology, began in an Italian university in the 1920s (Azzi 1928; Bensin 1925). Since then, agroecology education in colleges and universities of developed countries has grown in stature in response to the problems of industrial agriculture and capitalized food systems. There remains an opportunity to elevate the status of agroecology teaching and to break down the disparities between developed and developing counties.

1 Essential Elements of University Education for Agricultural Sustainability

Our planet is at risk because of overexploitation by humans. Therefore, sustainable development (SD) was proposed as an ideal approach to humanity's necessary and imperative transition (WCED 1987). However, to realize Sustainable Development Goals (SDGs), technology or engineering alone is not enough; there needs to be an integration of natural and social sciences to create institutions that effectively and systematically harness sustainability (Cash et al. 2003; Berg 2018). This, in turn, requires a fundamental paradigm transformation in the current education system, particularly in higher education, changing fragmented disciplines into holistic and connected disciplines and putting 'connection with nature', 'interdisciplinary collaboration,' and 'critical thinking' as core principles (Science Editorial Broad 2018; Loyalka et al. 2021).

Fundamentally, the knowledge system for SD constitutes a triangle, i.e. communication, translation, and mediation (Cash et al. 2003):

- **Communication:** Active, iterative, and inclusive communication among the stakeholders, experts, and decision-makers are the prerequisite to probing into and solving the problems of SD. Investigating the individual details of and the relationships among economic, environmental, and societal aspects requires communication, in particular between the natural and social sciences. The willing and fluent use of two-way dialogue among experts in natural and social sciences is very difficult but urgently needed.

- **Translation:** Once we achieve effective communication between experts and decision-makers, the results need to be translated into action by policymakers. The formation of an action agenda which meets the needs of stakeholders is the next step in approaching sustainability. By doing so, it is necessary to acknowledge that there is a long way to go from theory to application. Finding appropriate languages and procedures to translate policy into an action agenda is the priority.
- **Mediation:** There is another triangle to be rationalized in terms of meeting the objectives of 'policy to agenda', i.e. legitimacy, credibility, and salience. The concept of legitimacy in this context manifests as the impartial consideration and treatment of conflicting viewpoints and interests. Credibility, however, pertains to the scientific soundness and sufficiency of the technical evidence and the arguments put forth. Salience, meanwhile, is concerned with the degree to which the assessment aligns with and addresses the requirements of the decision-makers. Mediating the conflicts among them is based on understanding the three pillars of economic, environmental, and social sustainability.

The abovementioned processes approaching sustainability will focus on competence or competency for sustainable development fostered or nurtured in schools (Barth et al. 2007). Competence is defined as 'a combination of skills, abilities, and knowledge needed to perform a specific task' by the US Department of Education. In 'Four Pillars of Education for the 21st Century' for approaching sustainable development, competence was expressed in cliched terms as 'learning to know, learning to do, learning to be, and learning to live together' and was grouped linguistically into 'domain competences (professional capacity), methodological competencies (cognitive and creative capacity), personal competences (personal virtues connecting to cultural consciousness), and social competences (intersubjective communication ability)' (Mochizuki and Fadeeva 2010).

Many experts suggest that the creation of a learning society is the first mission of education towards the pursuit of sustainability. In doing so, cultural consciousness in valuing the 'Need' in the concept of SD is crucially important, and competence development is also value laden as well, which depends on people and their organizations' needs at the time.

Both because of agriculture's immense contributions and pressing challenges to the world population, agricultural sustainability in concept and sustainable agriculture in practice were put on the schedule right after SD was defined (Huffman 1990; Tilman et al. 2002). Competence development for agricultural sustainability and capacity training for sustainable agriculture should be the core mission of higher education in Agriculture-Associated Universities and Colleges (AAUCs) but remains challenging (Galt et al. 2013). As a good case, Galt et al. (2013) created affective learning ladders, i.e. Ways of Knowing and Learning; Understanding Values; The Inquiry Process; Analysis; Interpersonal Skills; Writing; and Presenting (Table 7.1), in parallel with Bloom's cognitive learning ladder in their course 'Community and Regional Development (CRD) 20: Food Systems'. Later, Bloom (1956) proposed six ladders for the students acquiring knowledge, i.e. memorizing,

Table 7.1 Learning ladders for the affective domain competence towards agri-food system sustainability (adapted from Galt et al. 2013, used with permission)

Competencies	Index
Ways of knowing and learning	• Understand your preferred learning style • Develop ideas for improving your individual and collaborative learning • Reflect on learning experiences, as an individual and as a group, in the classroom and in the field • Identify differences between various epistemologies (a.k.a. research perspectives, theoretical lenses)
Understanding values	• Examine personal values as they relate to food and the food system • Identify values underlying peer and popular conceptions of sustainable food and food systems • Explain the differences between one's own values and the values of others concerning the sustainability of food and food systems
The inquiry process	• Pose research questions that address your interests and correspond to a theoretical lens used in agri-food studies • Create interview questions to answer research questions • Use field-based research methods (interviews and observation) involving food system actors and locations • Perform collaborative learning in teams through field research and analysis • Identify gaps in current knowledge and perspective
Analysis	• Identify sections of the food system (production, processing, distribution, retail, consumption, and disposal) • Compare and contrast segments of the food system (conventional, organic, etc.) • Assess multiple locations within the food system using questions related to social, economic, and environmental criteria • Demonstrate systems thinking, including identification of components, relations, and setting boundaries • Interpret food choices based on different cultural identities and positions in society • Describe the relationship between structure and agency • Operationalize theoretical lens used in agri-food studies
Interpersonal skills	• Understand small-group processes • Practice group decision-making through dialogue and consensus • Co-manage fieldwork project logistics
Writing	• Organize and express ideas clearly in outline and/or draft form • Connect concepts and ideas from class, labs, readings/videos, and your own ideas • Engage in the revision process as a necessary part of good writing
Presenting	• Demonstrate understanding of principles of public speaking • Present research findings using visual, oral, and textual communication

comprehension, application, analysis, evaluation and creation (or synthesis), which could be grouped into two phases: basic and advanced knowledge transfer (Roiling and de Jong 1998). Through the completion of these two successive steps (ladders) (Fig. 7.1), students are empowered to cultivate the confidence, values, and forward-thinking vision necessary to forge ahead. They are then able to translate the knowledge they have acquired into practical action, specifically in the process of

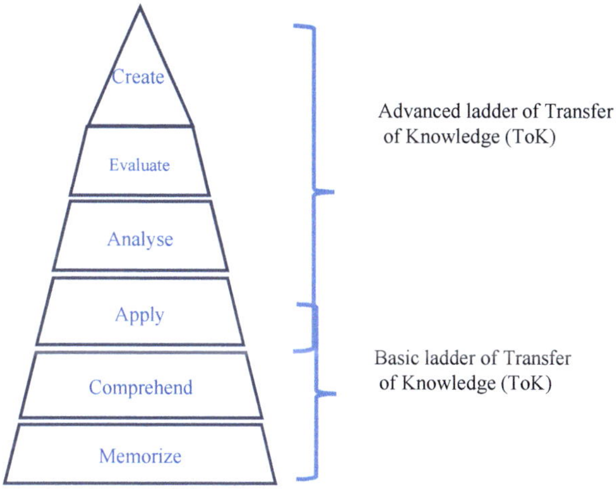

Fig. 7.1 The combined cognitive learning model of knowledge transfer

constructing an innovative model for the agri-food system within the context of higher education. This endeavour is ultimately geared towards the overarching goal of achieving agricultural sustainability (Fig. 7.2).

Such a cognitive learning model was also used in agroecology course designing and teaching (Lieblein et al. 2007), which will be addressed in the next section of this chapter.

Another outstanding case is from Migliorini and Lieblein (2016). The authors summarized the pedagogy of a sustainable agriculture course in Food Culture and Communication Master Program majoring in Human Ecology and Sustainability in Italian University of Gastronomic Sciences (UNISG). To enable the students in the program to integrate their prior knowledge with new knowledge through reflective experience enlightened by the famous educator John Dewey's experience learning paradigm (Dewey 1938, 1942, 2005), the course was designed into six cognitive learning ladders, i.e. observation, participation, dialogue, visioning, and reflection, in developing students' competences for SD:

> *Observation* is the competency of carefully examining situations in the 'world out there' with which you are confronted, before you make any judgements about the situation. This has the intention of an unbiased examination. *Participation* is the competency of participating in work in the field, not as a distant observer, but rather with personal commitment and dedication in interaction with different stakeholders. *Visioning* is the competency of not just repeating what already has been thought, but to think 'outside the box' in new and creative ways. It is about the ability to visualize a desired future so concretely that you would recognize it if you came across it. *Reflection* means the ability to link own experiences to theory in sustainable agriculture and to personal development. *Dialogue* is the core competency of performing a two-way communication.

The self-evaluation index is listed in Table 7.2.

Such cognitive learning ladder was also used in agroecology course designing and teaching (Lieblein et al. 2012), which will be addressed in the next section of this chapter.

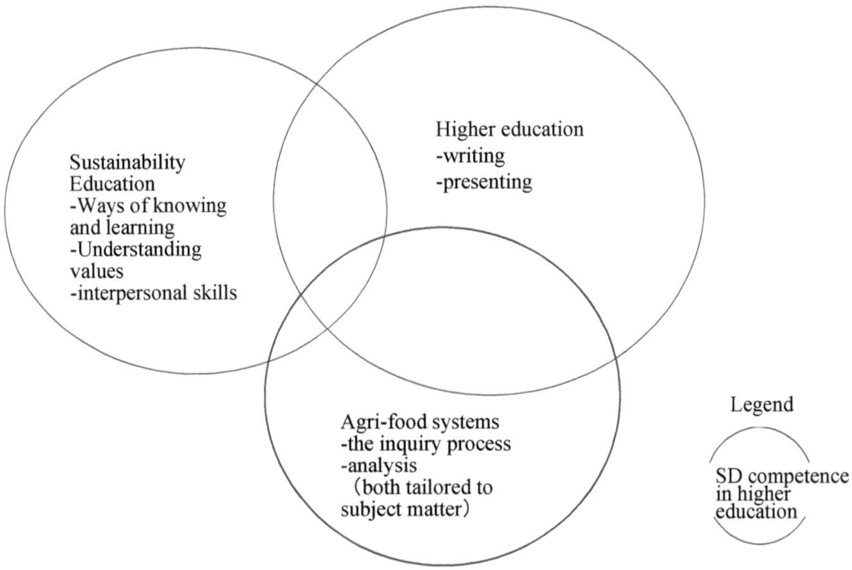

Fig. 7.2 SD competence fostering linking to agri-food systems. (Adapted from Galt et al. 2013, used with permission)

Table 7.2 The self-evaluation index of John Dewey's six ladder cognitive learning in terms of agri-food system education (adapted from Migliorini and Lieblein 2016, used with permission)

Competences	Index
Observation	• Carefully observe a situation in the field • Display such a situation as a rich picture • Allow for examination of the whole situation before drawing conclusions
Participation	• Recognize values and goal conflicts of different stakeholders in society • Participate in work 'out in the field' with commitment and dedication • Empathize with the goals and feelings of stakeholders in the field
Visioning	• Envision situations that do not yet exist • Articulate innovative ways of approaching challenges met by stakeholders in society
Reflection	• Knowledge about sustainable agriculture • Connect situations in the field to theory in sustainable agriculture as well as to personal growth • Self-guided learning • Awareness of the role of reflection in personal learning and development
Dialogue	• Understand and respect other people's views • Set formal roles and competencies aside when needed in a conversation • Distinguish between a debate and a dialogue

2 New Challenge for Education in Agroecology for Sustainability

Rather than some specific courses abstracted from agricultural and food systems aiming at sustainability, which to some degree still is confined in the scope of the fragmented disciplines and subjects in universities, agroecology, which has been defined as the sciences of sustainable agriculture (Gliessman 1990, 1997) and then as the ecology of the entire food system (Francis et al. 2003; Gliessman 2007), is more appropriate to be a holistic and a unifying science and discipline for guiding transformation to agricultural sustainability(Francis and Gliessman 2023). Within the disciplinary scope of agroecology, sustainability has been characterized as a key property of agroecosystem besides its basic focus on productivity by G. Conway in 1985 (Conway 1985):

> Sustainability is the ability to maintain a desired level of production over time in spite of long-term destabilizing influences. Sustainability is a measure of persistence or long-term durability of a production system.

Therefore, under the banner of agricultural sustainability as the core objective of agricultural transformation since the 1990s, some universities in North America started to build agroecology courses and academic programs, accordingly, the Department of Agronomy in those universities has altered the department names with specific names related to ecosystems and environmental components, such as the department of water sciences, soil sciences, plant sciences, crop sciences, environmental sciences, natural resources management, or two combinations, transforming production focus to environment and resource focus (Raun et al. 1998), or highlighting the marriage of ecology and agricultural sciences (Paul and Philip Robertson 1989; Jackson and Piper 1989).

From the data at the end of 2018, there were 71 agroecology or analogous programs and 58 agroecology-programmed student-hosting farms among the universities in the United States and there are similar programs concurrent in the rest of the world (David and Bell 2018). Among those initiatives, the University of Illinois (College of Agriculture) was one of the pioneers, who started its agroecology program with Agroecosystem Analysis as its core course in the late 1980s right after the SD concept was formally defined by WECD and Low Input Sustainable Agriculture (LISA) proposed as one of the alternative agriculture models for industrial agriculture (Schweitzer 1989; Gerber 1990). At the time, the pioneer faculties in the USA universities, particularly in land-grant universities' faculty of agriculture, realized that it was their urgent responsibility to restructure curriculum and pedagogy to coordinate agroecological transformation from industrialized agriculture (Lieblein et al. 2000a, b). Efraim Hernandez X., an agronomy professor at the National School of Agriculture in Chapingo (UACh), Mexico; Miguel A. Altieri, an agroecology professor at the University of California in Berkeley, USA; Steve Gliessman, an agroecology professor at the University of California in Santa Cruz, USA; Wu Zhiqiang, an agroecology professor at Fujian Agricultural College, China; Charles

Francis, an agroecology professor at the University of Nebraska, USA; and Luo Shiming, an agroecology professor at South China Agricultural University, China—these agroecology pioneers initiated agroecology courses and, in some cases, agroecology programs in their host universities and colleges, and the events below are worthy of being noted.

Agroecology Program in the *University of California in Santa Cruz* In 1981, Professor Steve Gliessman offered an agroecology course under the environmental studies undergraduate program. The agroecology program was founded with the primary goal to develop and monitor the indicators of agricultural sustainability. The students were encouraged to integrate natural, economic, and social factors in pursuing sustainable production agriculture, and the program and curriculum design move gradually closer to sustainable agriculture by defining agroecology as ecology of sustainable agriculture. After 2003's agroecology definition by Francis et al., the agroecology program and curriculum evolved into a focus on sustainable food systems locally, regionally, and globally (Francis et al. 2003; Francis and Gliessman 2023).

With time, more agroecology courses and programs were established in several USA universities and institutes not limited to land-grant and state-owned universities. According to Francis and Gliessman (2023), in 2023, the University of Vermont established the Institute of Agroecology, which is expected to play a crucial role in nurturing the upcoming cohort of agroecologists and change agents for the transformation of the food system. It accomplishes this through on-campus educational programs, including an undergraduate Bachelor of Science degree in agroecology. Additionally, it offers certificate and micro-certificate programs for postgraduate study in agroecology. The institute further features a fellowship program aimed at undergraduate students with an interest in agroecological research and extension. For instance, the Agroecology Extension Undergraduate Summer Research Fellowship Program is one such initiative. The program also supports informal education endeavours and provides continuous educational opportunities for the faculty and staff of the University of Vermont.

At almost the same time, the University of Wisconsin-Madison, Purdue University, University of California, Davis, University of Massachusetts at Amherst, University of Connecticut, University of New Hampshire, Delaware Valley University, and University of Hawaii at Manoa offered bachelor or master or both degrees in agroecology-related programs with various names, such as agroecology and sustainable agriculture (see https://www.collegefactual.com/majors/agriculture-ag-operations/agricultural-production/agroecology-and-sustainable-agriculture/rankings/top-ranked/. Accessed April 19, 2025); and University of Florida offered undergraduate, master, and Ph.D. degrees in agroecology at the same time (see https://agronomy.ifas.ufl.edu/students/online/agroecology/. Accessed April 19, 2025).

Agroecology Program and Curriculum in Nordic Regions In 1994, Professor Charles Francis was invited to a scholarly visit to the Nordic region (including

Norway, Sweden, Estonia, and Lithuania) universities in the fields of forestry, veterinary, and agricultural (the so-called Nova Universities) (Francis 2004; Francis and Gliessman 2023) to initiate the agroecology program and within which a series of agroecology courses from undergraduate to postgraduate (including master and Ph.D. level) were offered under a broader definition of agroecology as the ecology for sustainable agriculture (see the niche of agroecology defined by Gliessman 1990). The background of Nordic agroecology programs and a series of agroecology courses existed with difficulty inside the compartmentalized disciplinary education in AAUCs, resulting in fossil energy reliance agriculture in the Nordic region of the European continent (Lieblein et al. 1999, 2008; Francis et al. 2001). The goal of their excellent teaching work was to make a transition in discipline, research, and teaching paradigm of agriculture under agroecology (see Fig. 7.3).

- ***Discipline Changing from Compartmentalized to Systematic Focuses:*** *By experiential learning and action research of agroecology, the faculties and students naturally explored system view instead of mono-farming factors in Nordic context.*
- ***Methodology Integrating Between Natural and Social Sciences:*** *By highlighting agroecology's interdisciplinary properties, the faculties and students integrated qualitative and quantitative research methods into agriculture investigation and infused moral and value view into the Nordic regional food system.*
- ***Pedagogy Transforming from Faculty Concept Teaching to Student Action Learning:*** *By designing multiple landscape studies and bringing students to*

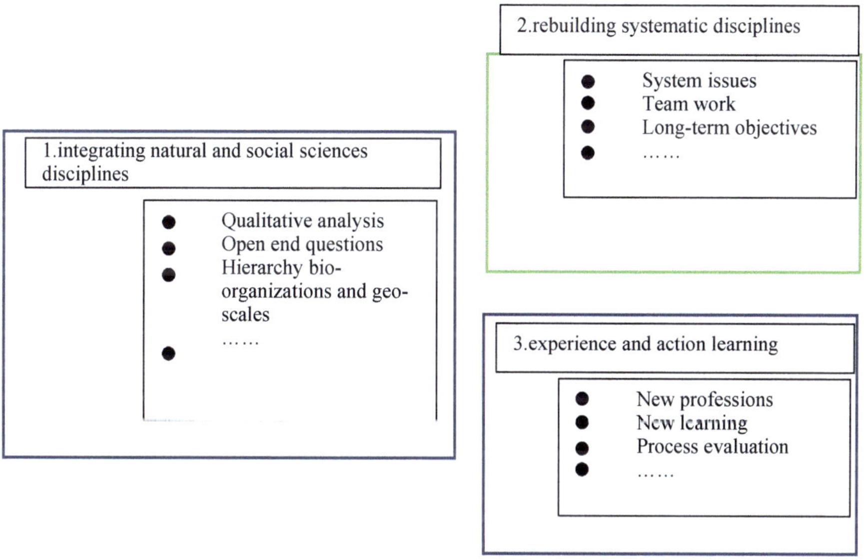

Fig. 7.3 Research and teaching paradigm transition via agroecology program and curriculum. (Revised from Charles A. Francis et al. 2001)

fields, faculties gave up their classroom authority to immerse students in the Nordic real-world context.

Among them, a Ph.D. ecological agriculture course and a master's agroecology program were recorded in detail (Lieblein et al. 1999, 2008, see Boxes 7.1 and 7.2).

Box 7.1 Fostering Ecological Agriculture Competence Through Ph.D. Courses (Abstracted from Lieblein et al. 1999, 2008)

- **Course Background**
  The majority of the existing research initiatives and educational curricula in Nordic countries are structured around themes that are rather narrow in scope and have a short-term focus. In contrast, agroecological transformation involves the consideration of entire systems and aims to ensure their long-term sustainability
- **Course Objectives**
  For a Ph.D. student aiming to independently carry out research in the fields of agroecology, agroecosystem, or ecological agriculture, it is essential to possess a solid foundation in scientific principles and experimental methodologies. Additionally, such a student should be equipped with an array of other research techniques that enable in-depth inquiry across various levels of the spatial hierarchy.
- **Course Description**
  In 1995, an ecological agriculture course was formulated for Ph.D. students enrolled in the agroecology program and was implemented across universities in the Nordic/Baltic Region. The course structure consisted of several key elements: a pre-course literature review accompanied by a written thesis summary, a week-long immersive workshop held at Fokhol Farm in Stange, Norway, and subsequent follow-through reports and evaluations. To ensure active participation, a variety of instructional approaches were employed. Lectures, large group discussions, and small-group working sessions were utilized to integrate each student into a team, with the guidance of facilitators. This setup enabled students to engage with the challenges associated with research design and implementation.
  Moreover, the course presented traditional methods of experiment design, both in research station settings and on farms. In addition to these, different social science methods were also introduced, along with the concepts and methodologies of systems.
- **Learning Outcomes**

  To acquire knowledge of and engage in the practical application of a diverse array of inquiry methodologies for the purpose of studying and assessing not only individual component technologies but also the more comprehensive aspects of practices, farms, and food systems.

(continued)

Box 7.1 (continued)

To gain an understanding of the appropriate timing and contexts in which these methods should be implemented, as well as the capacity to incorporate these concepts and techniques into the design of their own Ph.D. projects.

To acknowledge the significance of various hierarchical levels at which questions can be posed, and most participants embraced certain novel methods that enhanced the relevance of their research.

To broaden the network of scientific colleagues with whom students can seek advice and consultation as they progress in their research careers.

Box 7.2 Nordic Agroecology MSc Program for Future Sustainable Agri-Food System (Information Abstracted from Lieblein et al. 2004, 2008)

- **Program Background**
  Being aware that food production, national and local food security, and equity in access to adequate nutrition are of paramount importance is merely the initial step. In the context of developing sustainable food systems on a global scale, two pivotal issues come to the fore. The first is education regarding the methodology of conducting research in this field, and the second is the dissemination of information to the general public to enhance their awareness of the food, nutrition, and hunger challenges confronting society.
- **Program Goals**
  To equip contemporary students in the Nordic region with the skills and motivation to become driven researchers, proficient communicators, and catalysts for change in future systems poses a substantial challenge via enhancing education in agroecology.
  To serve these master's students as enablers of the transition towards sustainability within our food systems by formulating an agroecological pedagogy suitable for the Nordic region.
- **Program Description**
  The Agroecology Program that confers the Master of Science (MSc) degree was initiated in 1995. Its primary objective is to comprehensively examine the production, economic, environmental, and social aspects of farming and food systems within the Nordic region. The agroecology courses within this program are structured as a 2-year curriculum divided into two distinct stages, progressing from the study of farming systems to food systems. An action learning and holistic systems approach is employed, resulting in the development of an 8-week module centred on

(continued)

Box 7.2 (continued)

farming systems, followed by a second 8-week module dedicated to food systems. These modules are designed based on the prior experiences of both the faculty members and the students.

In each stage of the coursework, an experiential learning and action research strategy is implemented. Students are actively engaged in spending time on farms and within communities, interacting with key stakeholders who play diverse roles within the farming and food systems. This immersive experience allows students to gain practical insights and a deeper understanding of the complex dynamics at play.

After completing two additional semesters of courses in their specific areas of interest, each student is required to undertake and successfully complete a thesis project. This project focuses on exploring and addressing specific aspects related to agroecology and the sustainability of food systems, thereby demonstrating their ability to apply the knowledge and skills acquired throughout the program.

- **Program Outcomes**

  Integrating teaching and research has proven to be extremely effective in equipping agroecologists to manage the complexity and uncertainty related to future food systems and nutrition. Firstly, it enables them to handle the system as open and interacting. Secondly, it helps them establish their focus at a systems level on farms and in the wider landscape. Thirdly, it encourages them to look at systems in an interdisciplinary way, integrating natural and social sciences. Finally, it promotes the use of experiential learning as a pedagogical base for understanding complex systems and action learning as the starting point in the field.

Sino-Canadian Co-teaching Agroecology Course In 1981, agroecology was introduced as a scientific discipline in China at the First National Conference of Agroecology and China's Agricultural Modernization. At the time, agricultural modernization proposed by China's government to pursue higher agricultural production characterized by mechanization, irrigation, chemical engineering, and 'electrification' was criticized greatly, and the concept of ecology, agroecology, and ecological agriculture was introduced at the conference. Since then, several agricultural colleges and universities, Northern China Agricultural University, Fujian Agriculture and Forestry University (FAFU), and Shenyang Agricultural University initiated agroecology courses for undergraduate students. At FAFU, the agroecology course was first offered to the horticultural science major in 1981 by Professor Wu Zhiqiang, then to agronomic science and soil science majors in 1982.

In 2004, the authors of this book introduced an agroecology course from the Nova Scotia Agricultural College (NSAC) in Canada for the students from the FAFU-NSAC 2 + 2 International Cooperative Education Programs, Horticulture, and Agricultural Resources & Environment as their mandatory subject. This course was taught using the Canadian pedagogical model, including such in-class techniques as hands-on, interactive learning, 'Think-Jot-Pair-Share', electronic learning platforms, online assignments and discussions, oral presentations, and after-class questionnaires, aimed at continuous active learning and critical thinking. This 21 years of experience (2004–2024) of co-teaching and curriculum development internationally in agroecology not only brought a huge pedagogic transformation in the classroom but also stimulated us, together with our graduates to put agroecological principles into building new agricultural models into practice in the Fujian region (Wang et al. 2019), which are summarized as follows (Wang 2023):

- *Given the existing long-term 'blind men and elephant' discipline differentiation of the agricultural curricula in China's agriculture-associated universities and colleges, this course constructs the macro-agricultural curricula to probe into the relationship between agricultural biology, agri-environment sciences, and social dimensions of agriculture to give students a systematic view of agriculture.*
- *Given the existing phenomenon that traditional agricultural courses in China's agriculture-associated universities and colleges are 'teacher-cantered' or 'faculty concept' and make students 'uneasy in class', the participatory learning or action learning of the course is fully implemented to let students devote themselves to the course learning.*
- *Given the existing traditional examination form of 'one examination determines the score' in the courses of China's agriculture-associated universities and colleges, students feel that 'credits are worthless', and adopting a 'diversified' learning effect evaluation system to encourage students to participate in the whole process of course learning.*

Following the Sino-Canadian agroecology co-taught course initiative, an international Curriculum Development in Agroecology involving several countries was established in 2023. Here is an example: the establishment of a master's program in agroecology, which is supported by the Erasmus+ initiative financed by the European Union, endeavours to enhance the capabilities of higher education institutions. This project is spearheaded by Hue University in Vietnam, collaborating with the Vietnam National University of Agriculture, Benguet State University (Philippines), Central Luzon State University (Philippines), Rajarata University (Sri Lanka), University of Peradeniya (Sri Lanka), Mendel University (Czech Republic), Polytechnic of Coimbra (Portugal), and Novel Group Sarl (Luxembourg). The endeavour of collaboration is to explore the social requirements for an agroecology training program in the partner countries with the closer interaction and co-production knowledge of the famer stakeholders in those countries (see https://www.agroecologyproject.eu/. Accessed April 19, 2025).

Besides the above initiative on agroecology education, Mexico in South America, Canada in North America, European Union, Outer Mongolia in East Asia University of Queensland in Australia, and University of São Paulo in Brazil offer undergraduate and master programs named sustainable agriculture and agroecology or agroecology and sustainable agriculture, pioneering in agroecology education.

- **Agroecology Education in Mexico**: According to Francis et al. (2003) and Francis (2004), an Agroecology course (Agroecosistemas de Mexico) has been part of the curriculum in Mexico's agricultural college since 1976, offered by Efraim Hernandez X., Faculty of the National School of Agriculture in Chapingo (UACh); after then, a series of courses named Ecology, Agroecology, and Tropical Agroecosystems were offered in the Ecology Department of the Colegio Superior de Agricultura Tropical (CSAT) in Tabasco, Mexico, which highlighted the values of traditional agroecosystems, and agroecology in a cultural context was provided by the Graduate Program in Anthropology at the Universidad Iberoamericana in Mexico City. Following a very successful intensive short course in tropical agroecology in 1989 at the Colegio de Postgraduados in Tabasco (the former CSAT), the Department of Agroecology was initiated with a full Agroecology curriculum. In addition, Agroecology programs were started in the Postgraduate School in Chapingo and University of the Yucatan in 1991. This latter initiative was stimulated by an international symposium on agroecology held there in 1990 which emphasized ecologically based production processes and practice. To date, more than 130 students have graduated and become catalysts for even more agroecology programs and curricula throughout Mexico.
    - **Agroecology Education in Canada**: Almost at the same time in 1995, the University of British Columbia (UBC) and the University of Manitoba (UM) of Canada set up an agroecology program for undergraduates in their Faculty of Agricultural Sciences, respectively. The program description on UBC website is:

> Agroecology is one of the four unique undergraduate programs offered through the Faculty of Agricultural Sciences at the University of British Columbia. Other undergraduate programs include Community & Environment Global Resource Systems Food, and Nutrition & Health. These programs bring together the elements of agricultural sciences, ecology and environmental thought and is influenced by the experiences of people who manage land and water to produce food and other products. The program provides an agricultural education that integrates disciplinary knowledge within a framework of ecological principles and encourages students to tailor their learning experiences to prepare for a wide range of careers with varying emphases, including animal studies, horticulture, soils and environment, resource economics and general agroecology.

- In the University of Manitoba, the Bachelor of Science in Agroecology enables students to experience the means by which we can attain agricultural production that is economically sustainable, all the while conserving resources and preserving the integrity of our ecosystems (see: https://umanitoba.ca/explore/programs-of-study/agroecology-bsc, accessed August 25, 2024). This is still in operation but the number of first year students has been declining.
- However, UBC changed the agroecology program name to applied biology due to the limited student's numbers and they altered the faculty name to Faculty of Land and Food Systems. Fortunately, the three-level courses of Agroecology, Agroecology I (Introduction to Agroecology, course code:

APBI 260) for the second-grade student, Agroecology II (course code: APBI 360) for the third-grade student, and Agroecology III (Advanced Agroecology, course code: APBI 460) for the fourth-grade students within the Applied Biology program are still being taught. The UBC farm near campus still functions as the site of action learning of agroecology.

- **Agroecology Education in Europe**: Including agroecology programs in the Nordic region, agroecology programs and courses are distributed in many universities and institutions in the European continent (Table 7.3, information abstracted from Wezel et al. 2018 and Francis and Gliessman 2023), showing that much attention is being paid to agroecology in the European Union to convert unsustainably industrial agriculture to sustainably ecological agriculture (Poux and Aubert 2018).
- **Agroecology Education in Outer Mongolia**: In Outer Mongolia, there are currently two agroecology programs for undergraduates at the Mongolian University of Life Sciences (https://www.muls.edu.mn/) and the University (school) of Agroecology and Business (http://ipas.edu.mn/en/admission/why-school-of-agroecology-and-business). The former is accommodated by the university's Faculty of Agroecology, and the latter is embedded in the university name, which is unique in the world.

Although there has been a substantial effort to address current challenges of agri-food system education globally, and many agroecology programs and courses have been offered to enhance student understanding of the complexity of agroecosystems towards agricultural sustainability, there remain two huge challenges in agroecology education worldwide (Wang et al. 2019):

- ***Building the Substantial and Applicable Connection Between Concepts of Agricultural Sustainability and Agroecology***: Although more attention has been paid to agroecology, and more papers on agroecology have been published under the broad recognition of the limitation of reductionist disciplines and sciences towards the ideal of agricultural sustainability in universities and colleges (Fiore and Borrello 2024), until now the theory of agroecology is mostly transplanted directly from general ecology without its specific theories and principles, particularly it lacks substantial integration with the related disciplines such as economics and social sciences. Moreover, teaching methods in agroecology have tended to be more didactic than practical and analytical; students and tutors have not consistently connected theory with practice in terms of agricultural sustainability or sustainable agriculture. Meanwhile, the pedagogy of agroecology in developing countries has lagged that of developed countries in teaching agroecology in which experience and action oriented teaching and research systems have been popularized (Bell and Bellon 2018; David and Bell 2018). Agroecology should have occupied a prominent position in the academic system of AAUCs of developing countries like NOVA universities from undergraduate to postgraduate (Jordan et al. 2005; Lieblein et al. 2008, 2012; Francis and Porter 2011); however, it is hardly realized in developing countries with the exception of several agroecology programs set in Outer Mongolian universities.

Table 7.3 Academic agroecology programs in European universities and institutions

Level	Name	Universities and institutions	Language
Master	Agroecology and Food Security (from Sept. 2018 onwards Agroecology, Water and Food Sovereignty)	University of Coventry (United Kingdom)	English
Master	European Master of Agroecology (double degree)	ISARA, Lyon (France), Norwegian University of Life Science, NMBU (Norway),	English
Master	Agroecology—Organic Agriculture (double degree)	ISARA, Lyon (France), Wageningen University (The Netherlands)	English
Master	Agroecology	Norwegian University of Life Science, NMBU, Aas (Norway)	English
Master	Agroecology	Swedish University of Agricultural Sciences (Sweden)	English
Master	Agroecology	Harper Adams University (United Kingdom)	English
Master	Master Interuniversitaire en Agroécologie	University of Gembloux (Belgium), Free University of Brussels (Belgium), University of Paris-Saclay, AgroParisTech (France)	French
Master	Agroecologia: Un Enfoque para la Sustentabilidad Rural	International University of Andalusia (Spain), Pablo de Olavide University (Spain), University of Cordoba (Spain)	Spanish
Bachelor, Master	Agroecology	University of Zagreb (Croatia)	Croatian
Bachelor, Master	Agroecology	Mendel University, Brno (Czech Republic)	Czech
Bachelor, Master	Study in Agricultural Ecosystems and Agroecology	University of South Bohemia (Czech Republic)	Czech
Bachelor, Master	Agroecology	**Swedish University of Agricultural Sciences (SLU)**	English
Summer School	International Summer School in Agroecology	ENSFEA and University of Toulouse (France), SLU (Sweden), Pablo de Olavide University (Spain), University of Liège (Belgium)	English
Summer School	Agroecology Summer School	ISARA, Lyon (France)	English
Summer School	Agroecology: small-scale and sustainable cultivation in theory and practice	Karlskoga Folkhögskola (Sweden)	Swedish
Bachelor	Agroecology	**Wageningen University & Research**	English
Summer School	Agroecology and animal production	Agreenium (France), Wageningen University (The Netherlands)	English

(continued)

Table 7.3 (continued)

Level	Name	Universities and institutions	Language
Ph.D.	Agroecology	Aarhus University (Denmark)	English
Ph.D.	Agroecology	University of Coventry (UK)	English
Inter-University Certificate	Agroécologie et transition vers des systèmes alimentaires durables	Université catholique de Louvain, l'Université de Liège, l'Université libre de Bruxelles	French
MOOC	Université Virtuelle d'Agroécologie	INRA, Agreenium, ISARA (France)	French
Master	Agroecology	**University of Copenhagen**	English

- ***Extending and Expanding the Vision and Agroecology Towards the Ecology of the Whole Agri-food Value Chains*:** Like the increase in scientific papers on agroecology, agroecology textbooks have also flourished. To our understanding, the content of agroecology textbooks still rests on the general narratives of the biological scales, following the ecological hierarchies of 'individual-population-community-ecosystem' routes, lacking the fulfilment of the desired goals to measure and reflect the agricultural sustainability of agroecology as an independent and holistic discipline. Particularly, when conducting a comparative study between several outstanding agroecology textbooks by Stephen Griessman and other various textbooks from developing countries, it is clear the former shows the characteristics of 'ecology in agriculture' with less comprehensive teaching and less focus on the acquisition of practical skills by the students and the latter exhibits the characteristics of the 'ecology of agriculture', showing more agroecology itself and integrating theoretical and practical teaching in whole. Accordingly, our recommendations for overcoming the challenges are:
- ***Applying Systematic Approaches to Formulate a Comprehensive Discipline of Agroecology:*** To develop both a theoretically sound and an applicably practical framework of agroecology towards agricultural sustainability within the agroecology program in AAUCs, one needs first to understand the scope of agroecology integrating the eco-socioecological sustainability of agricultural sustainability. To do this, it is necessary to integrate the theoretical, methodological, and applied dimensions of agroecology into a whole system (Table 7.4). The student must explore organism and environment, the two components of an agroecosystem as its basic units of theoretical dimensions to form the complete spectrum of agroecology. Based on this system, an agroecology course and curriculum of the agroecology program are constructed in AAUCs.

- ***Characterizing Bio-organizational Scale to Develop a Series of Subjects of Agroecology:*** To develop an independent discipline of agroecology from its parental disciplines of ecology and biology, as well as to develop the curriculum of the agroecology program in AAUCs, there is a strong need to broaden the framework of agroecology by enhanced integration of study subjects. Herein, we present the proposed subject family of agroecology embracing several key subsubjects (Table 7.4), among which 'Introduction to Agroecology' serves as a

Table 7.4 The proposed system-oriented teaching model for agroecology (from Wang et al. 2019, used with permission)

Study hierarchies	Study themes	Objectives
Theoretical system of subject		
Organism subsystem		
Above ecosystem	Biosphere security	To safeguard security of life on Earth
Ecosystem	Patterns and management of agroecosystem	To ensure food security and safety, fair trade, and distributing benefits and welfare in impartiality and equilibrium
	Agro-landscape theory and resilience of fragile eco-zone	To set up environmentally friendly, biodiversity-based agroecosystem
	To set up an environmentally friendly, biodiversity-based agroecosystem	To supply integrated technologies for resource management, exploitation, and maintenance
Community	Structure and function of field ecosystem: bio-niche and poly-cropping theories	To establish the multi-agroforestry agroecosystem based on the biodiversity theory
	Agrobiodiversity theories, evaluation, and exploitation	To set up the biocontrol system of disease, insect and weed control by means of crop diversity
Population	Production ecology: agricultural climate for yield formation	To attain high yield, quality, and efficiency
	Population structure on production quality	To address the mechanism of suitable population structure in agriculture
	Chemical ecology in field: crop allelopathy and its utilization	To clarify the certain allelopathic effects between various crops
Individual	Chemical ecology in field: crop allelopathy and its utilization	To clarify the certain allelopathic effects between various crops
	Rules and regulation mechanism of individual crops	To develop the culturing technologies in different growth stages
	Physiological ecology: individual and ecological factors	To analyse the effect of single and restricted combinations of environmental factors on physiological tolerance and resistance
Cell under individual	Plant molecular ecology: crop quality and resistance formation	To investigate the cell and molecular determinants of crop yield
	GMO: benefits and problems	To establish the evaluation framework and index for GMO being employed in the food system in the long term
	'Green gene': from crop farming to molecular farming	To utilize the so-called green gene methods and expand their application to the field
Environmental subsystem		

(continued)

Table 7.4 (continued)

Study hierarchies	Study themes	Objectives
Global agroecosystem	Global climate change: effect on crop production	To simulate the climate change impact on crop production local and global
Regional agroecosystem	Rule of resource utilization and environment protection	To differentiate regional resources and develop different utilizing strategies
Production environment	Soil, atmosphere, and water: quantity and quality	To clarify the relationship between the micro-climate and agricultural organism growth
Ecological factor	Stress ecology: dynamic and adaptation	To address effects of the altered single factor on the individual organism and its adaptability
Applied system of subject		
Ideology	Environmentally friendly and ecologically compatible	To review the concept of development and resource management in agriculture
Theoretical strategy	Sustainable development	To establish the sustainable agriculture system
Pattern selection	Alternative agricultures	To select patterns among ecological agriculture, organic agriculture, etc., according to natural and social conditions of nation
Technological integration	Agro-eco-engineers and technologies	To set up a precise agriculture system taking advantage of the biological, ecological, and information technologies, as well as agricultural equipment
	Genetic engineering on phytoremediation of the field pollution	To excavate 'green gene' to enhance the yield and quality of resistance of crops by means of genetic engineering
	Pest integrated management (IPM)	To develop IPM strategies to decrease chemical input and thus safeguard the food safety
Methodological system of subject		
Systematic analysis and modelling		To supply effective ways of management using system modelling
Energy and material flow analysis		To evaluate an agroecosystem by investigating its energy, matter, information, and value flow
Practical system of the subject (course)		
Field investigation and survey		To investigate the structure and characteristics of specific agroecosystem
Mathematic simulation and induction		To predict growing trends of certain organisms by combination of field and lab data
Field and lab experiments		To test the data with lab and field trials

bridge between ecology, agronomy, and rural sociology. 'Agroecosystem management' uses a holistic ecosystem-based approach for sustainable management of natural resources and the environment within an agroecosystem. 'Agroecological Economics' balances the ecological and economic factors for developing sustainable agriculture. 'Bio-farming Systems' integrates agrobiodiversity and healthy soil tillage based on principles of agroecology. 'Production Ecology' addresses the theory of crop production and its regulating methodology; 'Landscape Agroecology' deals with spatial aspects of agriculture; 'Engineering Agroecology' bridges the theory application of Agroecology towards designing an sustainable agroecosystem. 'Rural Ecology' aims to unscramble rural sustainability.'Political Agroecology' makes an institutional arrangement for a sustainable food system and its relevant education and research aspects of agroecology. Finally, the subject 'Field and Laboratory Investigations in Agroecology' aims to link research with education and help the students acquire relevant practical skills. Other themes can be added to the list such as crop ecophysiology, information agroecology, molecular farming, sustainable agriculture, and ecological agriculture.

3 Towards the General and Liberal Education in AAUCs Bridging with Agroecology

The current challenges for AAUCs are multiple: firstly, due to lower recognition of agroecosystem health and services based on current free-market economics, it is difficult to attract qualified students devoted to the conceptual rebuilding of the agriculture and food system (Lieblein et al. 2007; Loconto et al. 2018). Secondly, due to some historical disciplinary boundaries in course and curriculum, there are some barriers to integration of the discrete and fragmented knowledge presently taught at AAUCs (Francis et al. 2004, 2020); thirdly, to prepare the students for the immediate challenges they will face when they graduate, there is a need to move towards more student centred teaching and learning, with built in experiential learning and action research experience (Wiedenhoeft et al. 2003; Francis and Porter 2011; Migliorini and Lieblein 2016).

The key to the new science and education of agriculture and food system transformation in AAUCs is finding a sweet spot between economy, environment, and health. The concept can be a useful vehicle for approaching the realization of the SDGs; it is a simple visual tool that can help in deliberations around analysing a production input, a crop, a cropping system, a farming system, or even a value chain. The key question is whether whatever is being done produces a good balance of economics, environment, and human health. Agriculture and food are at the centre of human health (Power 2010; O'Rourke et al. 2017). No matter what system we have for food production, it is essential to have balance to be truly sustainable. Sustainability of farming systems has traditionally been defined as one where economics and the environment are balanced. For long-term sustainability of agroecosystems, in which humans are central to the success, it is essential that we add a third corner, health, without it everything will be in vain, the so-called one health concept (Francis and Gliessman 2023) to the balance loop.

To acquire such balance, there is a need to stimulate informed coordinated action through educating the next generation of agroecologists (Lieblein et al. 2000b, 2004; Francis et al. 2017). The recognition of the necessity to stimulate new informed thinking across cultures and nations, to use the fundamentals of agroecology for novel approaches and positive action, formed the basis for our Sino-Canadian joint agroecology education initiative over two decades (see Sect. 2).

In the past two decades, the emerging discipline of agroecology, initially defined as a science but now with a much larger social and economic footprint, offers the template for AAUCs' educational and pedagogical transformation. This feeds directly into the realization of the SDGs. China, the largest developing nation (when it comes to SDGs, perhaps one should consider all countries as developing) initiated a rural revitalization strategy in 2017 (Wang 2019). This strategy recognizes the complexity of the agriculture and food systems that is the basis of rural areas. This innate agricultural infrastructure provides an arena for positive application of agroecology, both defined as an interdisciplinary science, a set of practices, and a social movement (Wezel et al. 2009; Gliessman 2018; Loconto and Fouilleux 2019). This approach can also help inform AAUC's transformation in the global South.

Based upon recent experience of agroecological research, teaching, and community engagement in both developed and developing nations, the achievement of those development goals is found in the 'Sweet Spot' at the confluence between economy, environment, and society. Agricultural value chains need to be addressed as cooperative endeavours with twin goals of hunger and poverty alleviation and health and wealth improvements. The closer connection between SDGs and agroecology creates tremendous needs for the educational and pedagogical transformation in AAUCs in the global South.

3.1 *New Agriculture Needs New Approaches to Science and Education*

Despite the advocacy for an integrated education system in many universities, current university education is predominantly based on a speciality or major in response to the dominant reductionist model (Lieblein et al. 1999; Francis and Carter 2001; Salomonsson et al. 2005, 2009). The curriculum and courses are generally designed to cover the information in specific portions of the established disciplines (Francis and Altieri 1992), which usually focus on the knowledge pieces without the opportunity for integration in a practical way before graduation. Improved integration of diverse specialties would facilitate easier transition into real-world ecological problems (Francis 2010). An agroecological transition imperative needs a change from discrete disciplines to integrative methods, holistic thinking, and problem-based teaching and learning, a shift in focus from teacher to student and from teaching to learning, taking more of a global approach which would assist students as future change agents (Nicot et al. 2018).

The roles that AAUCs should play under SDGs and rural revitalization in global South include:

- Developing a holistic, integrative discipline, and science from the present compartmentalized, specialized knowledge approach, an eco-systematic and interdisciplinary approach to form new curriculum systems. Confronted with the major problems and challenges that modern agriculture has produced, there are fundamental reasons to rethink our current agricultural methods. Since the second world war, our basic ideology has been embedded in controlling nature and in solving problems with technologies that intervene in nature's functions. This approach has caused both ecological and social disruption. Therefore, as Wendell Berry once said, 'we need a science of agriculture that is authentically new—a science that freely and generously accepts the farm, the local ecosystem, and the local community as contexts, and then devotes itself to the relationship between farming and its ecological and cultural supports' (Kirschenmann 2004). Agroecology, linking ecology to the sciences of agriculture to form the holistic view of agriculture and food system, has been recognized as one core science to reach the SDGs. In short, the goal is to develop the student becoming an agroecologist, an expert both with innovative skills and knowledge (Lieblein et al. 2004; Francis et al. 2017).
- Enhancing learner's capability in undergraduate science programs through experiential and action-oriented teaching and learning. Action-oriented teaching and learning enable students to become agroecologists from study and practical experience, rather than passive learning in hierarchical educational systems (mostly classroom lectures) (Lieblein et al. 2004; Alaniz and Ri 2019; Francis et al. 2020). Educators with practical knowledge can improve the teaching environment, because they integrate theory into practice effectively and adjust course content based on new challenges and practical experiences to which they were exposed (Asebo 2019), in turn, enhancing students' abilities to apply theoretical frameworks to practice with effectiveness and confidence (Wrenn and Wrenn 2009). Educators should create phenomenon-based agroecology learning opportunities for students (Østergaard et al. 2010; Francis et al. 2013). These opportunities include field trips, dialogue with stakeholders, working on farms and in food systems, having a reflection and appreciation of the whole system on farms and the full value chain (Bacon et al. 2011), promoting collaborations between student learning with farmer and industry partners. At the university level, there is a need to identify funds to support additional resources and facilities allowing for practical and hands-on learning activities.
- Educating today's students to become motivated researchers, effective communicators, and agents of change in future systems. Conducting projects on farms, group activities, community engagement, and an experiential learning approach

stimulate and motivating students to become facilitators working towards sustainable agriculture (Lieblein et al. 2008). Educators can facilitate students to identify their learning needs, encourage students to ask questions, be creative and develop their own views, as well as to provide opportunities in learning through sharing experiences, such as small-group discussions and presentations (Wrenn and Wrenn 2009).

3.2 Agroecology: An integration of the Skills, Knowledge, Attitudes, and Visioning for the Graduates to Become a Better Agroecologist

Current education in AAUCs has concentrated on learning skills and knowledge based on agricultural expertise of a specific major. Knowledge and skills are essential to complete professional work in agriculture; graduates need knowledge about plants and animals in agriculture and soil sciences, but knowing the interactions among plant, animal, and soil will make for a better agroecologist who can have a deeper insight into utilizing the agroecosystem appropriately. Therefore, attitudes and visioning towards agriculture, farmers, and rural land territories (i.e. agroecosystem) are very important. In this regard, agroecology programs and courses need to embrace the following points to foster future agroecologists (Lieblein et al. 2007; Francis et al. 2017).

- Educational programs will develop autonomous and social learners.
- Graduate study in agroecology prepares students for responsible action in the future.
- Close working relationships with farmers and other stakeholders are essential to education.
- Local food systems provide an alternative to growing globalization of food.
- Ecological principles are applied in design of locally effective farming systems.

Those capacities of teaching agroecology are illustrated as follows (Lieblein et al. 2007; Francis et al. 2017; Mendez et al. 2017).

- **Observation**: Due to the complexity of agroecosystem and food systems in the global context, everyone involved with agroecology learning and teaching should be identified as an observer, and the instructors act as facilitators in the field. Hence, observation is our first task at the beginning of the course, rather than presenting opinion and theories.
- **Dialogue**: The process of observation provides an opportunity for dialogue by using facilitation based on the principles of agroecology as a platform to form a learning community among the instructors and students (learners and audience). Teachers recognize students' experience as an indispensable element in team learning, and students acknowledge teacher's expertise in helping them become agroecologists.

- **Participation**: Following distinguished educator John Dewey's 'learning by doing' principle, agroecology instructors design the participatory processes for students' responsible action in agroecology teaching and learning processes, including such approaches as classroom and online discussion, field survey, and presentations.
- **Reflection**: Following a period of team learning, in which students have embraced the diverse experiences resulting from participation, discussions, field experience, and survey, they are asked to present their ideas and thinking on a specific agroecosystem and food system and to share their reflection.
- **Visioning**: Based on students' skills, knowledge, capacities, and learning attitudes accumulated in the agroecology team and experiential learning process, they can envision an optimistic future in rebuilding sustainable agriculture and food system, even the rural ecosystem.

Therefore, experiential learning is critical for students to obtain competencies in agriculture and sustainability (Migliorini and Lieblein 2016). Such a learning approach provides students the opportunities of observation, participation, visioning, interaction, and reflection (Migliorini and Lieblein 2016). Learning from direct experience has been suggested as a strategy to motivate and introduce students to current challenges in agricultural production and food. Students benefit greatly from having real-word experience by working on farms, communicating with stakeholders in the full value chain, and learning from people within the food system (Waldenström et al. 2008; Østergaard et al. 2010).

References

Alaniz, E. P., and N. M. D. Ri. 2019. Education, work and vocational training in the MST: the experience of the Course of Agroecology of the Milton Santos School. *Revista Brasileira de Educação do Campo* 4:e6789.

Azzi, G. 1928. *Agricultural ecology (in Italian), Edition*, 237. Turin: Tipografia Editrice Torinese.

Bacon, C. M., D. Mulvaney, T. B. Ball, E. M. DuPuis, S. R. Gliessman, et al. 2011. The creation of an integrated sustainability curriculum and student praxis projects. *International Journal of Sustainability in Higher Education* 12 (2): 193–208.

Barth, Matthias, Jasmin Godemann, Marco Rieckmann, and Ute Stoltenberg. 2007. Developing key competencies for sustainable development in higher education. *International Journal of Sustainability in Higher Education* 8 (4): 416–430.

Bell, M. M., and S. Bellon. 2018. Generalization without universalization: towards an agroecology theory. *Agroecology and Sustainable Food Systems* 42 (6): 605–611.

Bensin, B. M. 1925. Agroecological characteristics description and classification of the local corn varieties chorotypes. Book, (Publisher unknown so far, B. M. Bensin was born in Russia on 30 January 1881. He received his education in agricultural sciences at the University of Minnesota, graduating with an M.S. in 1912).

Berg, Jemey. 2018. Tomorrow's earth. *Science* 360 (6396): 1379.

Bloom, B. S. 1956. *Taxonomy of educational objectives: the classification of educational goals*. New York: Longmans, Green.

Cash, David W., William C. Clark, Frank Alcock, Nancy M. Dickson, Noelle Eckley, David H. Guston, Jill Jager, and Ronald B. Mitchell. 2003. Knowledge systems for sustainable development. *PNAS* 100 (14): 8086–8091.

Conway, G. 1985. Agroecosystem analysis. *Agricultural Administration* 20:31–55.
David, C., and M. M. Bell. 2018. New challenges for education in agroecology. *Agroecology and Sustainable Food Systems* 42 (6): 612–619.
Dewey, J. 1938. *Experience & education*. New York: Collier Books.
Dewey, J. 1942. *Democracy and education*. New York: Macmillan.
Dewey, J. 2005. *Art as experience*. London: Penguin Books.
Fiore, Vincenzo, Massimiliano Borrello, Domenico Carlucci, Giacomo Giannoccaro, Simone Russo, Sarah Stempfle, and Luigi Roselli. 2024. The socio-economic issues of agroecology: a scoping review. *Agricultural and Food Economics* 12:16.
Francis, C. A. 2004. Education in agroecology and integrated systems. *Journal of Crop Improvement* 11 (1/2): 21–43.
Francis, C. A. 2010. Conventional research on controversial issues: an exercise in futility? *Renewable Agriculture and Food Systems* 25:3–7.
Francis, C. A., and M. A. Altieri. 1992. Agroecology and sustainable development—innovative ideas for an effective university curriculum. *Journal of Sustainable Agriculture* 3:107–112.
Francis, Charles, and Stephen Gliessman. 2023. Teaching agroecology: preparing students for navigating uncharted territory. *Agroecology and Sustainable Food Systems* 47 (10): 1431–1439. https://doi.org/10.1080/21683565.2023.2253186.
Francis, C. A., and P. Porter. 2011. Ecology in sustainable agriculture practices and systems. *Critical Reviews in Plant Sciences* 30:64–73.
Francis, C. A., G. Lieblein, J. Helenius, L. Slomonsson, H. Olsen, and J. Porter. 2001. Challenges in designing ecological agriculture education: a Nordic perspective on change. *American Journal of Alternative Agriculture* 16 (2): 89–95.
Francis, Charles A., and H. C. Carter. 2001. Participatory education for sustainable agriculture: everyone a teacher, everyone a learner. *ournal of Sustainable Agriculture* 18 (1): 71–83.
Francis, C., G. Lieblein, S. Gliessman, T. A. Breland, N. Creamer, Harwood, L. Salomonsson, J. Helenius, D. Rickerl, R. Salvador, M. Wiedenhoeft, S. Simmons, P. Allen, M. Altieri, C. Flora, and R. Poincelot. 2003. Agroecology: the ecology of food systems. *Journal of Sustainable Agriculture* 22:99–118.
Francis, C., T. A. Breland, E. Ostergaard, G. Lieblein, and S. Morse. 2013. Phenomenon-based learning in agroecology: a prerequisite for Transdisciplinarity and responsible action. *Agroecology and Sustainable Food Systems* 37:60–75.
Francis, C. A., E. S. Jensen, G. Lieblein, and T. A. Breland. 2017. Agroecologist education for sustainable development of farming and food systems. *Agronomy Journal* 109:23–32.
Francis, C. A., A. M. Nicolaysen, G. Lieblein, and T. A. Breland. 2020. Transformative education in agroecology: student, teacher, and client involvement in co-learning. *International Journal of Agriculture and Nature Resources* 47:280–294.
Galt, Ryan E., Damian Parr, and Janaki Jagannath. 2013. Facilitating competency development in sustainable agriculture and food systems education: a self-assessment approach. *International Journal of Agricultural Sustainability* 11 (1): 69–88.
Gerber, John M. 1990. *Agricultural sustainability and the University of Illinois: an introduction to agroecology*. The Agroecology Program Paper Reports of The University of Illinois, Urbana, IL, USA.
Gliessman, S. R. 1990. *Agroecology: researching the ecological basis for sustainable agriculture*, Ecological Studies Series No. 78. New York: Springer.
Gliessman, S. R. 1997. *Agroecology: ecological processes in sustainable agriculture*, 384. Boca Raton, FL: CRC Press.
Gliessman, S. R. 2007. *Agroecology: the ecology of sustainable food systems*, 384. New York: CRC Press, Taylor & Francis.
Gliessman, S. 2018. Defining agroecology. *Agroecology and Sustainable Food Systems* 42:599–600.
Huffman, E. 1990. Methods of assessing agricultural sustainability. *Environmental Monitoring and Assessment* 15:303–305.

Jackson, Wes, and Jon Piper. 1989. The necessity of marriage between ecology and agriculture. *Ecology* 70 (6): 1591–1593.

Jordan, N. R., D. A. Andow, and K. L. Mercer. 2005. New concepts in agroecology: a service-learning course. *Journal of Natural Resources and Life Sciences Education* 34:83–89.

Kirschenmann, K. 2004. *A new science for a new agriculture*. A paper prepared for the Mountain Sky Guest Ranch conference, "Evidence Based, Opinion Based and Real World Agriculture and Medicine," October 10–15, 2004.

Lieblein, G., C. A. Francis, L. Salomonsson, and N. Sriskandarajah. 1999. Ecological agriculture research: increasing competence through PhD courses. *The Journal of Agricultural Education and Extension* 6 (1): 31–46.

Lieblein, G., C. Francis, and J. King. 2000a. Conceptual framework for structuring future agricultural colleges and universities in industrial countries. *The Journal of Agricultural Education and Extension* 6 (4): 213–222.

Lieblein, G., C. Francis, W. Barth-Eide, H. Torjusen, S. Solberg, et al. 2000b. Future education in ecological agriculture and food systems: a student-faculty evaluation and planning process. *Journal of Sustainable Agriculture* 16 (4): 49–69.

Lieblein, G., E. Østergaard, and C. Francis. 2004. Becoming an agroecologist through action education. *International Journal of Agricultural Sustainability* 2:147–153.

Lieblein, G., T. A. Breland, E. Østergaard, L. Salomonsson, and C. Francis. 2007. Educational perspectives in agroecology: steps on a dual learning ladder toward responsible action. *NACTA Journal* 51:37–44.

Lieblein, G., T. A. Breland, L. Salomonsson, N. Sriskandarajah, and C. A. Francis. 2008. Educating tomorrow's agents of change for sustainable food systems: nordic agroecology MSc program. *Journal of Hunger & Environmental Nutrition* 3 (2–3): 309–327.

Lieblein, G., T. A. Breland, C. Francis, and E. Ostergaard. 2012. Agroecology education: action-oriented learning and research. *The Journal of Agricultural Education and Extension* 18 (1): 27–40.

Loconto, A., and E. Fouilleux. 2019. Defining agroecology: exploring the circulation of knowledge in FAO's Global Dialogue. *International Journal of Sociology of Agriculture & Food* 25 (2): 116–137.

Loconto, A., A. Jimenez, E. Vandecandelaere, and F. Tartanac. 2018. Agroecology, local food systems and their markets. *Ager-Revista De Estudios Sobre Despoblacion Y Desarrollo Rural*13–42.

Loyalka, Prashant, Ou Lydia Liu, Guirong Li, Elena Kardanova, Igor Chirikov, Shangfeng Hu, Ningning Yu, Liping Ma, Fei Guo, Tara Beteille, Namrata Tognatta, Lin Gu, Guangming Ling, Denis Federiakin, Huan Wang, Saurabh Khanna, Ashutosh Bhuradia, Zhaolei Shi, and Yanyan Li. 2021. Skill levels and gains in university STEM education in China, India, Russia and the United States. *Nature Human Behaviour* 5 (7): 892–904. https://doi.org/10.1038/s41562-021-01062-3.

Mendez, V. E., M. Caswell, S. R. Gliessman, and R. Cohen. 2017. Integrating agroecology and participatory action research (PAR): lessons from Central America. *Sustainability* 9 (705): 1–19.

Migliorini, Paola, and Geir Lieblein. 2016. Facilitating transformation and competence development in sustainable agriculture university education: an experiential and action oriented approach. *Sustainability* 8:1243.

Mochizuki, Yoko, and Zinaida Fadeeva. 2010. Competences for sustainable development and sustainability: significance and challenges for ESD. *International Journal of Sustainability in Higher Education* 11 (4): 391–403.

Nicot, R., S. Bellon, A. Loconto, and G. Ollivier. 2018. The European networks of research, education and training stakeholders in agroecology. *Open Agriculture* 3:537–552.

O'Rourke, M. E., M. S. DeLonge, and R. Salvador. 2017. Insights from agroecology and a critical next step: integrating human health. *Agroecology and Sustainable Food Systems* 41:880–884.

Østergaard, E., G. Lieblein, T. A. Breland, and C. A. Francis. 2010. Students learning agroecology: phenomsenon-based education for responsible action. *The Journal of Agricultural Education and Extension* 16:23–37.

Paul, Eldor A., and G. Philip Robertson. 1989. Ecology and agricultural sciences: a false dichotomy. *Ecology* 70 (6): 1594–1597.

Poux, Xavier, and Pierre-Marie Aubert. 2018. *An agroecological Europe in 2050: multifunctional agriculture for healthy eating*. Findings from the Ten Years For Agroecology (TYFA) modelling exercise, Iddri-AScA, Study N°09/18, Paris, France, p 74.

Power, A. G. 2010. Ecosystem services and agriculture: tradeoffs and synergies. *Philosophical Transactions of the Royal Society B-Biological Sciences* 365 (1554): 2959–2971.

Raun, W. R., N. T. Basta, J. A. Hattey, H. Zhang, and G. V. Johnson. 1998. Changing departmental names from agronomy to plant, crop, and soil sciences. *Journal of Natural Resources and Life Sciences Education* 27:113–116.

Rogers, A. 1996. *Participatory training: using critical reflection on experience in agricultural extension training*. FAO: Training for Agriculture and Rural Development, Economic and Social Development Series (54).

Roiling, N., and F. de Jong. 1998. Learning: shifting paradigms in education and extension studies. *Journal of Agricultural Extension & Education* 5 (3): 143–161.

Salomonsson, L., C. A. Francis, Lieblein, and B. Furugren. 2005. Just in time education. *NACTA Journal* 49:5–13.

Salomonsson, L., A. Nilsson, S. Palmer, A. Roigart, and C. Francis. 2009. Farming systems education: case study of Swedish test pilots. *Renewable Agriculture and Food Systems* 24:48–59.

Schweitzer, H.J. 1989. Sustainability and research thrusts for the University of Illinois. In *Fifteenth Annual Illinois Crop Protection Workshop*. March 7–9, 1989. Champaign, IL, USA.

Science Editorial Broad. 2018. Education for the future. *Science* 360 (6369): 1409–1412.

Tilman, David, Kenneth G. Cassman, Pamela A. Matson, Rosamond Naylor, and Stephen Polasky. 2002. Agricultural sustainability and intensive production practices. *Nature* 418:671–677.

Waldenström, Cecilia, Lennart Salomonsson, Charles Francis, Mike Moulton, and Geir Lieblein. 2008. Individualized student-centred education: prototype for an agroecology BSc programme. *International Journal of Agricultural Sustainability* 6 (4): 236–247. https://doi.org/10.3763/ijas.2008.0246.

Wang, Songliang. 2019. Concerted development of ecological agriculture along with community-supported agriculture to facilitate rural vitalization (in Chinese). *Chinese Journal of Eco-Agriculture* 27 (2): 212–217.

Wang, Songliang. 2023. The demand for the agriculture associated universities and colleges' teaching &learning transformation and educational evaluation reformation under the rural revitalization strategy: exemplified by agroecology course teaching. *China Agricultural Education* 24 (3): 8–16.

Wang, S. L., C. D. Caldwell, S. L. Kilyanek, and Sean M. Smukler. 2019. Using agroecology to stimulate the greening of agriculture in China: a reflection on 15 years of teaching and curriculum development. *International Journal of Agricultural Sustainability* 4:298–311.

Wezel, A., J. Goette, E. Lagneaux, G. Passuello, E. Reisman, RodierC., and G. Turpin. 2018. Agroecology in Europe: research, education, collective action networks, and alternative food systems. *Sustainability* 10:1214.

Wezel, Alexander, S. Bellon, T. Doré, C. Francis, D. Vallod, and C. David. 2009. Agroecology as a science, a movement and a practice. A review. *Agronomy for Sustainable Development* 29 (4): 503–515.Wiedenhoeft, M., S. Simmons, R. Salvador, and G. McAndrews. 2003. Agroecosystems analysis from the grass roots: a multidimensional experiential learning course. *Journal of Natural Resources and Life Sciences Education* 32:73.

World Commission on Environment and Development. 1987. *Our common future*. Oxford: Oxford University Press.

Wrenn, J., and B. Wrenn. 2009. Enhancing learning by integrating theory and practice. *The International Journal of Teaching and Learning in Higher Education* 21:258–265.

Chapter 8
Systematic and Participatory Action Research in Agroecology

Abstract Research in agroecology provides the theoretical framework for making the transition to sustainability in world agriculture and food systems. The last decade has seen an African advancement and interest in the integration of systematic and participatory action research in agroecology to scale both up and out agroecology to this end. This shows that the university-based research paradigm and stakeholder-oriented research methodology should be integrated in postgraduate-level research in agroecology to envision an applied outcome under the new vision of integrative indigenous knowledge.

Keywords Research in agroecology · Participatory action research · Indigenous knowledge · Ecological agriculture development · Food Sovereignty · Rural development

Learning Objectives

After reading this chapter, students should be able to:

1. Describe, with examples, university-based systematic action research in agroecology.
2. Explain, with examples, the importance of stakeholder-oriented participatory research in agroecology.
3. Link PAR with indigenous knowledge in agroecology research and practices.

****

You cannot solve a problem with the same thoughts that created it.
—Albert Einstein (Cited from Charles Francis et al. 2005)

Agricultural innovations respond better to local challenges when they are co-created through participatory processes.
—Food and Agriculture Organization of the United Nations (2018)

S. Wang, C. D. Caldwell, *Advanced Agroecology*,
https://doi.org/10.1007/978-3-032-08689-1_8

Agriculture in modern times, symbolized by Industrialized Agriculture, has been structured by the approach of segregated and specialized knowledge in universities and institutions. In the past six decades, huge progress in agricultural production has been achieved since the 1960s' first Green Revolution. However, this has been accompanied by environmental and social degradation, soil erosion and contamination, population explosion (particularly in developing countries), small-scale farmer poverty, and food safety problems. Consequently, there has been a call for a change (Sachet et al. 2021), which includes synthesizing disconnected knowledge into a coherent strategy for investigating the nature and problems of modern agriculture and then finding a way forward.

Starting in the 1980s, agroecology was able to offer a holistic view of agriculture, under which systematic thinking and approaches are used to examine the connection of components and diagnose the problems of agriculture. In this context, a systematic and participatory action research paradigm in agroecology has been advocated in universities and institutions since the 1990s and much progress has been made in this research paradigm revolution of agroecology.

Participatory research has a long history in agriculture. In the North American context, in the 1940s, 1950s, and 1960s, there was considerable farmer cooperative research done on site (International Institute for Environment and Development 1988). Historically, extension service workers would have close connection with many farmers, especially the innovators. It was known that working with the most innovative farmers was the best way of moving new methods and technology into the field, testing it, and making it common usage. There was recognition of the value of farmer knowledge. In North America, there are two excellent examples of participatory research with farmers:

- Extension Services: Agricultural extension services played a crucial role in participatory research 1945–1985. These services, often affiliated with land-grant universities, facilitated the exchange of knowledge between researchers and farmers. Extension agents worked closely with farmers to test and adapt new agricultural practices and technologies on their farms (International Institute for Environment and Development 1988). In Canada, the agricultural colleges were key in this effort.
- Soil Conservation Programs: In response to the Dust Bowl of the 1930s, the U.S. government-initiated soil conservation programs that continued into the 1940s and 1950s. These programs involved farmers in the development and implementation of soil conservation techniques, such as contour ploughing and crop rotation, to prevent soil erosion (International Institute for Environment and Development 1988). In a similar way, in Canada, there were conservation programs working closely with farmers.

These examples highlight the early efforts to involve farmers in the research process, laying the groundwork for more formalized participatory research methods in later decades. With the demise of many small agricultural colleges and the movement of agricultural education into larger universities, there was a movement away from the agriculture college approach to participatory research.

1 University-Based Systematic Action Research in Agroecology

Our world faces multiple challenges requiring paradigm shifts. Regarding education, according to interviews by *Science Journal* editorial with several famous scientists in the world, 'connection with nature' is the first one that education needs to change, 'interdisciplinary collaboration' ranks the second, and 'interdisciplinary collaboration' is on the list (Science Editorial Broad 2018). Of course, how to connect with nature, communicate, and collaborate with other disciplines and people have become core needs of humankind. Gaining these skills and attitudes in the education process are key to a research and education paradigm shift (Roiling and de Jong 1998). In recent decades, a revived paradigm in research called Action Research has gained much attention and made considerable progress in universities and institutions to enable learners to understand real questions in nature and society and then to explore workable solutions to those questions (Sendall et al. 2018).

According to Lieblein et al. (2008), Action Research (AR) originates from John Dewey's 'learning by doing' educational theory (Dewey 1916), and it was revived by Grundy (1982) and used for social and educational fields. AR is a research paradigm explicitly concerned with the improvement of situations through the taking of informed action and the development of relevant theory. It is then used to guide further action, which is characterized by improvement and involvement, but it is the idea of strategic action (Schubert 1995), which is deliberate and considered, action undertaken to bring about change, that distinguishes action research from other forms of action. It may be designed to assess the past or to find out about the present (Packham and Sriskandarajah 2005). Alternatively, AR may be an approach to gain insight into the present situation of the agroecosystem, to envision future wanted scenarios, and to accomplish desirable and workable outcomes (Carlsen et al. 2006). AR can be divided into systematic action research and participatory action research. In terms of looking at objects and people, it is a prerequisite of 'learning by doing', i.e. experiential learning, first proposed by David Kolb (1984). In this context, AR is inherently linked to action learning by mean of 'reflection' (Revans 1998; Mezirow 2000).

AR has three modes (Packham and Sriskandarajah 2005):

- **Technical Action Research**: This is a product-centred mode of AR that motivates the novel and efficient action 'idea' of facilitators and improves participants' practical skills.
- **Practical Action Research**: This mode aims to improve practice through the participants' wisdom. In doing this, a multibody participatory reflection process will be followed to gain action motivation and capacities.
- **Emancipatory Action Research:** This is a critique-oriented AR mode that follows the phases of theory, enlightenment, and action to reach the certain profound objectives of AR.

Agri-food systems, which connect with nature, with multiple stakeholders and with scientific disciplines, need action research and systematic approaches, and university and institutions, especially AAUCs, should be the pioneers in this field (see Chap. 7 of this book in detail). Fortunately, several universities in developed countries, which have been dominated by reductionist experimental research for about 150 years, took the paradigm shift by setting up integrative approach programs and organizations in agri-food system fields. The pioneering job was done by University of Western Sydney, Hawkesbury, in Australia in the 1980s (Bawden 1992). This was followed by the Midwest universities of the USA, such as Washington State University, the University of Idaho, the University of California, Davis, Texas A&M, Ohio State University, Iowa State University, and universities in Canada, such as University of Columbia (UBC). Since the 1990s, they gradually developed Systems Agriculture programs to address agri-food system issues. In the European Union, there are also several new research institutes and centres founded by changing from their traditional agronomic research paradigm in separated and fragmented disciplines to systematic approaches. Examples include the Centre for Sustainable Agriculture, the National Centre for Agroecology, and the Nordic Agroecology Program (see Chap. 7 of this book in detail), with the goal of nurturing the future agri-food system scientists who are not only good at their own area of expertise but also able to integrate other related disciplines into holistic thinking in tackling agri-food system problems (Bohlen and House 2009). Within those organizations and programs, action research is considered as instrumental to system-oriented problem-solving. This contrasts with the old approaches in which the problems are imagined in advance and solutions are concluded accordingly (Packham and Sriskandarajah 2005).

Agroecology was proposed as a discipline to utilize the principles of ecology to design and manage sustainable food systems (Francis et al. 2003; Gliessman 2007). A system typically represents multiple stakeholders, landscape scales, and academic disciplines. Researchers are devoted to addressing the real questions of food systems. Therefore, AR is a suitable means to approach the goal of agroecology as a discipline (Wezel et al. 2018).

Jordan et al. (2005) first incorporated action research and action learning in agroecology courses, considering that it is not sufficient for students to gain knowledge and skill based on their assumption of problems existing in agri-food system. Rather, students need to truly experience the problems of the farmers and managers within the systems. Therefore, instructors and students of agroecology courses are encouraged to set up research teams with agri-food stakeholders to approach the real problems and questions. Under such pedagogical ideology, action learning or experiential learning has been gradually incorporated into agroecology undergraduate curricula, and action research was incorporated in agroecology postgraduate curricula in the United States and the European Union (Lieblein et al. 1999, 2008; Wiedenhoeft et al. 2003). Table 8.1 shows examples from various publications.

Table 8.1 Action research and action learning in agroecology curriculum

Cases	Descriptions	References
1. Three continuous years' Ph.D. Agroecology courses in Norwegian University of Life Sciences	This international collaboration initiative attracted 10 Ph.D. students from 10 countries, which started to offer 8 days' Research Methods in Ecological Agriculture in 1995 aimed at integrating farming practices, crop rotations, crop/animal sciences, and marketing from farmer's views. Ph.D. student achieved the goal by interviewing farmers and experts working on the farms. The 1-week scheduled course Systems Research in Ecological Agriculture offered in 1996 aimed at understanding the whole food system questions through observing the four organic farms and interviewing the corresponding stakeholders and drew the conclusion by biological and social science integration. The third course proceeded in 1997 named the Spatial Spectrum from Farming Systems to Food Systems, which targeted at drawing the real questions of regional and global agri-food systems via the investigation into the farmers, processors, marketers, government officials, and consumer groups. As a distinguished result, the participants adjusted their questions for their doctoral research theme after the courses	Lieblein et al. (1999, 2008, 2012), Francis et al. (2001)
2. A Master Agroecology Program in Norwegian University of Life Sciences	Agroecology: from farming systems to food system, an MSC program in Nova University Network aiming at nurturing the agroecologists across European continent embarked in 1999 after the experience of Ph.D. agroecology course. The program started as action research on farm at the first semester to acquire the real questions on farm with periods of two 8-weeks agroecology courses, agroecology and farming system, agroecology, and food system interacting with producers and consumers, respectively, to gain real questions from 'just in time education' and evaluate with six ladder experience learning. Students selected their interesting topics from their host countries to conduct a thesis project in their second and third semesters. The program lasted 11 years ending in 2010, 189 students worldwide were enrolled and completed the 3 year's program with various understandings and skills as qualified agroecologists in agroecologically transformative farms all over the world	Francis et al. (2003, 2004, 2017), Lieblein et al. (2008, 2012)

(continued)

Table 8.1 (continued)

Cases	Descriptions	References
3. The Food From Thought graduate research assistantship program at the University of Guelph	Realizing that the common interest of action research and transdisciplinary research is to find real-world questions, the host (the Arrell Food Institute) of the Food From Thought graduate research assistantship program organizes four postgraduate students from different disciplines as an action research team, in cooperation with the non-academic organization to address real-world problems of community security, which results in the creation of a guidebook for Farmer Led Research (FLR) on community food security, which provides bridging and collaborating mechanisms (by providing the six workable pieces of advice for the transdisciplinary action research team) to overcome and to integrate the knowledge gaps between social and natural sciences, and writing style differences between various disciplines' researchers	Lam et al. (2019)
4. Systemic action research for postgraduate education in agriculture and rural development	Hawkesbury College of Advanced Education of the new University of Western Sydney was the first institute to set up a system agriculture problem in 1978, in which action research was initiated and considered a valuable method to address the system agriculture issues. The current project succeeds the system agriculture with researchers' learning experience to employ systematic action research to instruct Ph.D. research and education, consequently, by exemplifying 4 Ph.D. students' progress in finding their research hypotheses before and after the action research, indicating the system action research is a significant instrument to probe the read-world problems in facilitating sustainable agricultural and rural development	Packham and Sriskandarajah (2005)

2 Stakeholder-Oriented Participatory Action Research in Agroecology

The researchers engaging in agri-food systems need to embrace skills to involve all members of the food system to promote good communication and respect in the system to promote an equitable, sustainable system (Gliessman 2009). In this context, agroecology as a science to rebuild sustainable agriculture should prioritize holistic, participatory action research (Wezel et al. 2018). Besides the systematic action research for harnessing the holistic food system issues, participatory action research (PAR) has gradually become a promising method to realize the objective of transdisciplinary cooperation and integration in shaping a fair and transparent agri-food system process. According to Roiling and de Jong (1998), PAR consists of four kinds of research modes:

- ***Participatory Technology Development (PTD)*** *is embedded in farming system research* via *co-research, farmer research groups to identify locally adaptative and applicable technologies. By doing this, farmers become active researchers to solve the pressing problems they have identified by co-researching with experts from institutes.*
- ***Participatory Rural Appraisal (PRA)*** *contains a series of methods employed to identify and prioritize key problems and objectives together with the intended beneficiaries for rural development and planning* (Pretty et al. 1995).
- ***Participatory Learning and Action Research (PLAR)*** *designs the experiential learning project in farms with the participation of farmers by means of co-learning, farmers in the co-learning group follow the designed learning path to draw their conclusions from it. The typical example of PLAR is the prevailing worldwide Integrated Pest Management (IPM) Farmer Field Schools, which has gradually replaced the traditional Training and Visit System in plant protection technology extension processes.*
- ***Rapid Appraisal of Agricultural Knowledge Systems (RAAKS)*** *focuses on solving the problem by multiple actors and interdependent stakeholders of agri-food systems, the solution arises from the interactions of participants, actors, stakeholders, or others bridged by scientists and experts.*

Méndez et al. (2018) depicted and discussed two cases of PAR explored in Central American countries (Boxes 8.1 and 8.2).

Box 8.1 Agroecology and PAR with Smallholder Coffee Farmers in El Salvador

- **Background**
  Agroecology was originally coined as a discipline to integrate ecology with agronomy 100 years ago, currently, agroecology has developed a more transdisciplinary science between ecology, economics, agronomy, and sociology in the field of agriculture, food system, and rural landscape.

(continued)

Box 8.1 (continued)

As the ecology for entire food systems (Francis et al. 2003), the transdisciplinary characteristics of agroecology not only cross with academies but also cross with non-academic stakeholders of agri-food systems, among which farmers are a particularly important segment. Such characterization of agroecology provides an acting platform for PAR. The collaboration between agroecology and PAR has been prevalent in Central and South America where agroecology was considered in large extent as a social movement for facilitating food sovereignty, characterized by iterative cycles of research, reflection, and resulting in a series of outstanding achievements, as well as a great collection of experiences and lessons.

- **Case Descriptions**
  Starting with doctoral dissertation research in 1999 focusing on ecological agriculture in the municipality of Tacuba, a coffee-growing region of El Salvador, a profound PAR-agroecology collaboration event lasted until 2013. The actors of the event were involved with three smallholder coffee farmer cooperatives, participated with some non-coffee growers, and bridged by the diverse group of researchers and non-researchers. The objective was to verify whether agroecologically coffee planting could preserve biodiversity and maintain the householder's livelihood at the same time. However, with PAR integrating into research progress, which was grouped into three phases, i.e. building relationship and understanding the context (1999–2003); supporting action continuing research and sharing (2004–2007);and changing direction academic and action outcomes and process hiatus, the initiative objectives evolved as well with more and more actors 'participation'.
- **Outcomes**

  Several achievements were gained by agroecology-PAR integration:

  Sharing Research: With farmers' and other actors' participation and involvement, the goal of research on agroecology, biodiversity, and livelihood was redirected to real-world questions, in this case, what effects are the agroecologically transformative coffee expected to bring about? The long-term food security and economic sustainability. More sharing of the research brought more participation, thus more motivation for agroecological transition adoptions.
  Believing Collective Power: The Gordian knot in the agroecological transition among the scattered small-scale coffee farmers was overcome when they were involved in in-depth research that was directed to their problems, during the process, they showed more patience with the slower interest harvest, even with some degree of tolerance for temporary loss.
  Commitment to Participation: Farmers gradually elevated their focus and attention in PAR process from their fluctuating performances once

(continued)

Box 8.1 (continued)

they recognized that participation meant resource distribution, and equated it to empowerment. Their participation therefore becomes a farmer's habit and daily activity or commitment.

Humility: Humility is a key characteristic of multiple actors' decision-making process. Leader researchers set an example of humility at the beginning of the project to learn knowledge from other actors, particularly from farmers. The attribution of humility is gradually formed in the PAR and agroecology collaboration process when the actors gain mutual trust to approach the real-world questions of coffee agroecosystem and the favourable results.

Trust and Accountability: Trust and accountability are the primary reasons as well as the expected results of the PAR and agroecology integration. The results can only be acquired when the mutual trust between researchers and farmers and the cooperatives was established.

Communication: Effective communication produces constructive results. In this project, the lead researchers spent much time collecting local information and indigenous knowledge relating to the coffee agroecosystem and economic system, which provided a fundamental base for building an appropriately communicative mechanism throughout the PAR process, there were no barriers in resource distribution even when the farmers were not on site due to the trust and established communication mechanism.

Box 8.2 The Community Agroecology Network's (CAN) Agroecology and Participatory Knowledge Production and Exchange in Mexico and Nicaragua

- **Background**
  The Community Agroecology Network (CAN) is an international non-profit organization founded in 2000 in Mexico with the central commitment to apply agroecology for the sustainable rural community, agriculture, and food systems, covering the whole Central American regions, where coffee production and trade become an influential economic industry with ecological and social consequences. CAN initiates PAR and agroecology collaboration in the agri-food system for Central America by conducting action research on ecological agriculture and providing experiential courses of agroecology through its international network. The coffee case described here in Mexico and Nicaragua is one of them.
- **Case Description**
  The case began with a funding project, the Youth Leadership and Education for Sustainable Agriculture and Food Sovereignty Project(YLFS) in

(continued)

Box 8.2 (continued)

collaboration with CAN's national and international partners, the Union of Cooperatives of San Ramon (UCA San Ramon) in Nicaragua, and the NGO Vinculación y Desarrollo Agroecológico en el Café/Agroecological Networks and Development in Coffee (VIDA) in Mexico. The project integrated agroecology and PAR to facilitate collaborations between 234 coffee farming families crossing 12 coffee-growing communities to alleviate food insecurity and seasonal hunger in the project regions. This was highlighted by the Youth and Women Empowerment in a 'Sustainable Agriculture (vegetable-focused home gardens) Fund', which aims at improving household dietary diversity and diversifying women's income, and 'Women's Unpaid Labour (the development of AgroEco® coffee) fund', which aims at recognizing women's reproductive and productive labour in coffee production.

- **Outcomes**

  Several achievements were gained by CAN's agroecology-PAR integration in the project regions:

  Sharing Research: CAN and its partners discovered the evidence-based problems of regional agri-food systems by sharing interests and findings with stakeholders and actors to develop an actionable agenda through open dialogues and discussions.

  Believing in Collective Power: By bringing together the stakeholders and actors and approaching differentiating each group's interests and ways of knowing, we can consolidate the collective action on building regional sustainable food systems.

  Commitment to Participation: CAN acts as a learning organization to let farmers and their families share and devote to experiencing learning from each platform to democratize knowledge production processes and make the sciences and technology driven by integration workable.

  Humility: CAN guides the participants to have mild attitudes on the past imparity they have encountered and to optimistically look forward to the better future after the PAR-agroecology collaboration project.

  Trust and Accountability: CAN and its partners have had very accountable relationship with the involved stakeholders before the project was initiated, and the reflection with substantial outcomes at the end of the project indicated there existed an even better connection with trust and accountability between the participants.

  Communication. The 5 years of PAR circle, research, reflection, and action, imply that an effective and stronger communication mechanism has been built.

3 Linking PAR with Indigenous Knowledge in Agroecology Research and Practice

One of the outstanding properties of agroecology is that it is rooted in indigenous knowledge systems (IKS), i.e. ways of knowing and interacting with local people and landscapes that have deep connections to soil, land, plants, and animals (Méndez et al. 2018; Altieri et al. 2015). This is perhaps the best reason why agroecology research should be closely related to PAR (Zurayk 2001; Eksvärd 2010; Shahraki et al. 2023).

Indigenous knowledge is the knowledge created by the residents and landscape in contrast with the knowledge generated by universities and institutions (Warren 1991). Indigenous knowledge has been mostly overlooked by Western Experimental Science, on the pretext that it lacks scientific rigour and experimental data (Briggs 2005). Indigenous knowledge has come back into consideration since the public critique of modern economic concepts of development which stem from and are supported by a reductionist scientific paradigm.

Traditional agriculture was based upon indigenous (local) knowledge, but, over time, has turned to rely on modern reductionist sciences. This reliance has been questioned since the onset of the negative environmental impacts of agriculture which came into view in the 1960s. It was Allan in 1965 who brought back the issue of indigenous knowledge in agriculture to the modern academy (Allan 1965); since then, the academic arena has started to pay more attention to indigenous knowledge usually characterized by teachings regarding living in harmony with nature for sustainable development (Briggs 2005) and in agri-food system transformation (Zurayk 2001). For instance, land use patterns are one of the key factors affecting agroecosystems and climate change; hence, it is a pressing topic of agroecology. According to Zurayk (2001), a better land use pattern usually is usually reflected by indigenous traditions of land use prior to the use of modern technologies. This suggests that indigenous land and soil classification can be useful for land use planning. In doing this, the authors explored participatory action research (PAR), particularly participatory rural appraisal (PRA) linked to indigenous knowledge for land use planning in a semi-arid mountainous region of Lebanon. The results implied strongly that indigenous knowledge is an essential part of agroecology.

Such an integration between indigenous knowledge and ecological knowledge result in 'indigenous ecology', referred to 'alliances between environmental and indigenous social movement organizations to preserve and defend traditional ecological knowledge' (Carruthers 1996; Salami 2020). Described below are several cases that exemplify 'indigenous ecology' practices from various publications worldwide.

3.1 Agroforestry Ecosystem Development

An agroforestry ecosystem is one of the traditional agricultural systems containing significant agroecological concepts and bearing plentiful indigenous knowledge. On the one hand, incorporating indigenous knowledge into agroforestry development can facilitate farmer participation via PAR and, on the other hand, can enforce the collaboration between scientists and farmers to replenish indigenous knowledge to modern scientific knowledge. Coalescence serves to build a better agroforest ecosystem. Among the three PAR methods, participatory rural appraisal (PRA), as an essentially qualitative survey methodology, is often used to collect concrete indigenous knowledge from farmers (Walker et al. 1995). Together with other applicable and experiencing methods assembled into the Agroforestry Knowledge Toolkit (AKT), it provides a useful tool for integrating indigenous knowledge with the modern scientific understanding of agroforestry development (Thapa et al. 1995). A recent agroforestry development case in Bolivia showed that integration of indigenous knowledge with scientific knowledge in building agroforests could result in more agrobiodiversity and climate change adaptation (Jacobi et al. 2017). The integration is quite significant and gains the benefits and excludes disadvantages. To reach the goal, a participatory agroecology method was explored (Cockburn 2015).

Using indigenous knowledge to protect forests and biodiversity in Bah Buah village of Vietnam has been quite effective (Tarq K. Hassan 2015). In the villages of Yunnan Province, China, the villagers believe that forests of certain kinds and places around the village are sacred and should be preserved in any circumstance. They believe that those who cut and destroy the forest will be punished by disasters falling to their family. Hence such ritually indigenous knowledge maintains the forest and biodiversity under the forest, even the ecologically compatible agricultural pattern nearby (Fig. 8.1).

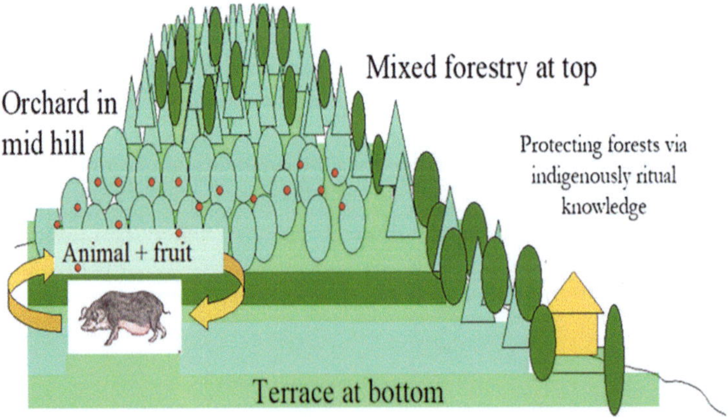

Fig. 8.1 Using indigenously ritual knowledge to protect forest helping neighbourhood eco-agriculture built in the village of Yunnan, China. (Courtesy: Prof. Luo Shiming, personal bestowing)

3.2 Nature Resource and Ecosystem Management

When dealing with natural resource management in agriculture, the combination of agroecological and indigenous knowledge demonstrates their advantages and strengths towards the trade-off between economic outcome and environmental benefits (Dewalt 1994). According to Dewalt (1994), modern knowledge systems of agriculture prefer to utilize external resources and energy such as machinery, fertilizer, and pesticides to maximize productivity and marketing return, resulting in matter and energy exhaustion, cultural losses and finally destruction of natural resources, jeopardizing the ecosystem. Indigenous knowledge, as another way of knowing nature and how agriculture interacts with scientific knowledge, might mitigate the destructive effects of modern agriculture and simultaneously produce ecological benefits that favour agriculture sustainability and natural resource utilization.

A typical case is managing crayfish resources in Lake Racken Watershed, in Sweden using local ecological knowledge (Olsson and Folke 2001). After communication with local fish farmers, the experts from institutions and NGOs adopted practices from local knowledge, including (Olsson and Folke 2001):

- Crayfish Harvesting Time: The local fishing association authorizes two separate days for fishermen to harvest crayfish, one at the beginning and one at the end of August.
- Crayfish Size Restriction: The minimum size of the crayfish is 10 cm which can be harvested for sustainable harvesting purposes.
- Number of Traps: Per household of fisherman is restricted to 15.

3.3 Ecological Agriculture Development

In the past 5 years, agroecology principles have been put in place for constructing eco-agriculture in Pingnan county of Fujian province of China. There were excellent experiences and lessons in integrating the two knowledge systems. At the beginning of the grassroots-based project in 2019, university professors, including one of the authors of this book, were entitled as experts and instructors, and new farmers as students, resulting in the project being stagnant. It was realized that the situation was not working and approaches were changed to involve the new farmer and members of cooperatives in formulating the agenda and decision-making processes by using the PAR method. Gradually, when experts became the bridge between farmers and rural-oriented NGOs, holding scientific and local knowledge, the eco-agriculture development project moved forward smoothly (Fig. 8.2).

In China, ecological agriculture was defined as using the principles and methods of agroecology to build both ecologically compatible and economically viable agriculture by applying technologies combining indigenous knowledge and modern scientific knowledge (Wang and Chen 2009). The eco-agriculture development case

Fig. 8.2 Academic experts communicating with local farmer in Qiantou Village of Pingnan County of Fujian Province on Sept. 16th, 2024 on eco-agriculture development. (Courtesy: Mr. Jixin Guo)

in Pingnan exhibits the processes of such knowledge integration and putting into practice.

3.4 Framing Food Sovereignty

Another important arena for integrating agroecology and indigenous knowledge is to frame food sovereignty because agroecology has been defined as a social movement for food sovereignty (see Chaps. 7 and 9). According to Maudrie et al. (2023), indigenous food sovereignty (IFS) is a holistic approach to food that incorporates values of relationality, reciprocity, and relationships, broadening the concept of food security. With this change, government department experts need to communicate with the indigenous people and absorb their understanding to form integration criteria for food security.

The case is from Vietnam described by Hassan (2015). In Dong Van district of Vietnam, located in a high rock mountain area, the farmers built 'pockets' in low-level land of the mountain and grew maize in the Pockets, namely 'rocky pocket agriculture', which helped secure the local food security. A similarly famous example of ancient China in the Pearl River Delta and the Taihu Lake Basin was 'Sangji Yutang' (or dyke-pond system) agroecosystem, which was created by the local farmers a thousand years ago exploring their local ecological knowledge (see Chap. 4). The example in Iran shows that using indigenous knowledge to choose the crop kind that adapts to the local environmental condition can result in food security via moulding food habits of local people (Hassan 2015).

In India, there is increasingly recognition of the problematic situation resulting from the application of the Western industrial agricultural model. In some parts of the country, such as Chuktia Bhunjia Tribe of Odisha, farmers are encouraged to explore the indigenous agricultural model that relied on indigenous knowledge ecologically, agronomically, and economically to safeguard the local food security and biodiversity (Sabar and Midya 2024). For instance, they refer to local agrometeorological knowledge to make cropping plans for adapting to climate change.

3.5 *Rural Development Planning and Implementation*

Rural development could be an indispensable part of societal development both geo-politically and eco-economically (Wang et al. 2021). This can be guided by ecological and agroecological knowledge generated by academia interacting with agroecological knowledge of rural people (Kmoch 2018).

Kmoch (2018) addressed the case of the participatory approach to collaborative research by Ar Yone Oo Social Development Association (Myanmar) and Chalmers University of Technology (Sweden) in Tedim and Tonzang Townships of northern Chin State and Kale Township of western Sagaing Region, Myanmar. It combined systematic thinking with participatory agroecological knowledge of farmers from 103 households among six sample villages (Box 8.3).

Box 8.3 Exploring Local Agroecological Knowledge for Rural Development

- **Background**
  The details and multiple sources of ecological knowledge for rural rebuilding after natural disasters and facing extreme weather under climate change remain a huge challenge and risk in rural Myanmar. Project organizations and planners need to find rigourous approaches for integrating scientific and local knowledge for rural development planning and implementation. In this context, 'Strengthening The Resilience Of Natural Disaster Affected Groups('STRONG)' project sponsored by the Federal Ministry for Economic Cooperation and Development of Germany (BMZ) was set up. And after 3 years' of project implementation, two international organizations, Ar Yone Oo Social Development Association(Myanmar) and Chalmers University of Technology (Sweden), conducted participatory research to address the lessons from the STRONG project.

(continued)

Box 8.3 (continued)

- **System thinking and participatory approaches**
  - System thinking explored aims at understanding the status of rural household livelihood status and its interacting external conditions such as social and ecological factors caused by the strong project intervention.
  - Participatory approaches such as brainstorming with STRONG project staff and stakeholders within project coverages to deduce the cause-effect chains in improving the resilience after the project.
- **Outcomes**

  Key insights from the participatory research with STRONG project stakeholders and participants are summarized as follows:

  - Collaborative research of STRONG project staff and academic experts interacting with rural people needs the time and commitment of each participant and results in the sharing and co-creation of knowledge.
  - Agroecological knowledge of rural people on adapting to climate change, farming system improvement, and interacting with expert knowledge of researchers have enhancing effects on the livelihood and food security of rural communities.
  - Participatory approaches such as brainstorming and causal-diagramming activities are well suited to knowledge co-creation and sharing processes for rural rebuilding after a disaster.
  - The reflection meeting is a key process of PAR in sharing and generating the integrating knowledge with participants.
  - Cause-effect chains resulting from the participatory research of rural development actors help to identify the effects of the STRONG project intervention.

In summation, indigenous knowledge has multifaceted functions in planning and implementing agri-food system and rural development in sustainable ways. This is because of indigenous knowledge's historical, geo-politically, and socio-ecological adaptation characteristics. When interacting with modern science and integrating into agroecology, indigenous knowledge provides a simple but effective pathway to sustainable development of agri-food system and rural landscapes. However, the state needs to make more effort to work with the local farmers and stakeholders of agri-food systems. By doing this, academic experts and agroecologists in this context need to bridge modern ecology and local agroecological knowledge for action research and education of agroecology to embody the farmers' subjectivity in securing food security and rural development sustainably.

References

Abdullah, Haidi Jamal, and Tarq K. Hassan. 2015. The use of indigenous knowledge in agriculture and its role in sustainable development. *International Journal of Scientific Research* 4 (10): 1312–1317.

Allan, W. 1965. *The African husbandman*. London: Oliver and Boyd.

Altieri, M. A., C. I. Nicholls, A. Henao, and M. A. Lana. 2015. Agroecology and the design of climate change-resilient farming systems. *Agronomy for Sustainable Development* 35 (3): 869–890.

Bawden, R. J. 1992. Systems approaches to agricultural development: the Hawkesbury experience. *Agricultural Systems* 40:153–176.

Bohlen, Patrick J., and Gar House. 2009. *Sustainable agroecosystem management: integrating ecology, economics and society*. Boca Raton, FL: CRC Press, Taylor & Francis Group.

Briggs, John. 2005. The use of indigenous knowledge in development: problems and challenges. *Progress in Development Studies* 5 (2): 99–114.

Carlsen, H.W., T.A. Breland, G. Lieblein, and C. Francis. 2006. *Developing organic farming in Norway through systemic action research*. Conference Poster at: Joint Organic Congress, Odense, Denmark, May 30–31, 2006. https://core.ac.uk/download/pdf/10923957.pdf. Accessed April 24, 2025.

Carruthers, David V. 1996. Indigenous ecology and the politics of linkage in Mexican social movements. *Third World Quarterly* 17 (5): 1007–1028.

Cockburn, J. 2015. Local knowledge/lacking knowledge: contradictions in participatory agroecology development in Bolivia. *Anthropologica* 57 (1): 169–183.

Dewalt, Billie R. 1994. Using indigenous knowledge to improve agriculture and natural resource management. *Human Organization* 53 (2): 123–131.

Dewey J. 1916. *Democracy and education: an introduction to the philosophy of education*. 1966 Edition. New York: Free Press.

Eksvärd, Karin. 2010. Facilitating systemic research and learning and the transition to agricultural sustainability. *The Journal of Agricultural Education and Extension* 16 (3): 265–280.

Food and Agriculture Organization of the United Nations. 2018. The 10 elements of agroecology. Guiding the transition to sustainable food and agricultural systems.

Francis, C. A., G. Lieblein, J. Helenius, L. Slomonsson, H. Olsen, and J. Porter. 2001. Challenges in designing ecological agriculture education: a Nordic perspective on change. *American Journal of Alternative Agriculture* 16 (2): 89–95.

Francis, C., G. Lieblein, S. Gliessman, T. A. Breland, N. Creamer, Salomonsson L. Harwood, J. Helenius, D. Rickerl, R. Salvador, M. Wiedenhoeft, S. Simmons, P. Allen, M. Altieri, C. Flora, and R. Poincelot. 2003. Agroecology: the ecology of food systems. *Journal of Sustainable Agriculture* 22:99–118.

Francis, Charles, Geir Lieblein, Havard Steinsholt, Tor Arvid Breland, Juha Helenius, Nadarajah Sriskandarajah, and Lennart Salomonsson. 2005. Food systems and environment: building positive rural-urban linkages. *Human Ecology Review* 12 (1): 60–71.

Francis, C. A., E. S. Jensen, G. Lieblein, and T. A. Breland. 2017. Agroecologist education for sustainable development of farming and food systems. *Agronomy Journal* 109:23–32.

Gliessman, S. R. 2007. *Agroecology: the ecology of sustainable food systems*, 384. New York: CRC Press, Taylor & Francis.

Gliessman, Steve. 2009. Editorial: community-based participatory action research with agroecology. *Journal of Sustainable Agriculture* 33:799–800.

Grundy, S. 1982. Three modes of action research. *Curriculum Perspectives* 2:23–34.

International Institute for Environment and Development (IIED). 1988. *Evolution of agricultural research and development since 1950: toward an integrated framework*. Gatekeeper Series No. 12 edited by Robert E. Rhoades. https://www.iied.org/sites/default/files/pdfs/migrate/6365IIED.pdf. Accessed April 21, 2025.

Jacobi, Johanna, Sarah-Lan Mathez-Stiefel, Helen Gambon, Stephan Rist, and Miguel Altieri. 2017. Whose knowledge, whose development? Use and role of local and external knowledge in agroforestry projects in Bolivia. *Environmental Management* 59:464–476.

Jordan, N., D. Andow, and K. Mercer. 2005. Ecology of agricultural systems: a service-learning course in agroecology. *Journal of Natural Resources and Life Sciences Education* 34:83–89.

Kmoch, Laura. 2018. *Addressing climate vulnerability and farming system challenges with local agroecological knowledge*. Technical report. https://research.chalmers.se/publication/506994/file/506994_Fulltext.pdf. Accessed September 14, 2024.

Kolb, D. 1984. *Experiential learning: experience as the source of learning and development*. Upper Saddle River, NJ: Prentice Hall.

Lam, Steven, Michelle Thompson, Kathleen Johnson, Cameron Fioret, and Sarah K. Hargreaves. 2019. Toward community food security through transdisciplinary action research. *Action Research* 19 (4): 1476–7503.

Lieblein, G., C. A. Francis, L. Salomonsson, and N. Sriskandarajah. 1999. Ecological agriculture research: increasing cosmpetence through PhD courses. *The Journal of Agricultural Education and Extension* 6 (1): 31–46.

Lieblein, G., T. A. Breland, L. Salomonsson, N. Sriskandarajah, and C. A. Francis. 2008. Educating tomorrow's agents of change for sustainable food systems: Nordic agroecology MSc program. *Journal of Hunger Environmental Nutrition* 3:309–327.

Lieblein, Geir, Tor Arvid Breland, Charles Francis, and Edvin Østergaard. 2012. Agroecology education: action-oriented learning and research. *The Journal of Agricultural Education and Extension* 18 (1): 27–40.

Maudrie, Tara L., Cassandra J. Nguyen, Rachel E. Wilbur, Megan Mucioki, Kaylee R. Clyma, Gary L. Ferguson, and Valarie Blue Bird Jernigan. 2023. Food security and food sovereignty: the difference between surviving and thriving. *Health Promotion Practice* 24 (6): 1075–1079.

Méndez, V. Ernesto, Martha Caswell, Stephen R. Gliessman, and Roseann Cohen. 2018. Integrating agroecology and participatory action research (PAR): lessons from Central America. *Sustainability* 9:705. https://doi.org/10.3390/su9050705.

Mezirow, J., ed. 2000. *Learning as transformation: critical perspectives on a theory in progress*. San Francisco, CA: Jossey-Bass.

Olsson, Per, and Carl Folke. 2001. Local ecological knowledge and institutional dynamics for ecosystem management: a study of Lake Racken Watershed, Sweden. *Ecosystems* 4:85–10.

Packham, Roger G., and Nadarajah Sriskandarajah. 2005. Systemic action research for postgraduate education in agriculture and rural development. *Systems Research and Behavioral Science* 22:119–130.

Pretty, J., I. Guijt, J. Thompson, and I. Scoones. 1995. *A trainer's guide for participatory learning and action*. London: NED.

Revans, R. 1998. *ABC of action learning*. London: Lemos and Crane.

Roiling, N., and F. de Jong. 1998. Learning: shifting paradigms in education and extension studies. *Journal of Agricultural Extension and Education* 5 (3): 143–161.

Sabar, Bhubaneswar, and Dipak K. Midya. 2024. Landscape: indigenous agriculture, sustainable food production and response to climate change—a case study of Chuktia Bhunjia Tribe of Odisha, India. *Journal of Asian and African Studies* 59 (1): 123–141.

Sachet, Erwan, Ole Mertz, Jean-François Le Coq, Gisella S. Cruz-Garcia, Wendy Francesconi, Muriel Bonin, and Marcela Quintero. 2021. Agroecological transitions: a systematic review of research approaches and prospects for participatory action methods. *Frontiers in Sustainable Food Systems* 5:709401. https://doi.org/10.3389/fsufs.2021.709401.

Salami, Rita 2020. The role of indigenous knowledge in sustainable urban agriculture and urban food security in Minna, Nigeria. *Libr Philos Pract* (e-journal). 3919. https://digitalcommons.unl.edu/libphilprac/3919.

Schubert, W. H. 1995. Students as action researchers: historical precedent and contradiction. *Curriculum and Teaching* 10 (2): 3–14.

Science Editorial Broad. 2018. Education for the future. *Science* 360 (6369): 1409–1412.

Sendall, Marguerite C., Laura K. McCosker, Alison Brodie, Melissa Hill, and Phil Crane. 2018. Participatory action research, mixed methods, and research teams: learning from philosophically juxtaposed methodologies for optimal research outcomes. *BMC Medical Research Methodology* 18:167. https://doi.org/10.1186/s12874-018-0636-1.

Shahraki, A. S., T. Panagopoulos, H. E. Ashari, and O. Bazrafshan. 2023. Relationship between indigenous knowledge development in agriculture and the sustainability of water resources. *Sustainability* 15:5665. https://doi.org/10.3390/su15075665.

Thapa, B., F. L. Sinclair, and D. H. Walker. 1995. Incorporation of indigenous knowledge and perspectives in agroforestry development. Part 2: case-study on the impact of explicit representation of farmers' knowledge. *Agroforestry Systems* 30 (1–2): 249–261.

Walker, D. H., E. L. Sinclair, and B. Thapa. 1995. Incorporation of indigenous knowledge and perspectives in agroforestry development part 1: review of methods and their application. *Agroforestry Systems* 30 (1–2): 235–248.

Wang, S. L., and D. M. Chen. 2009. Development of modern ecological agriculture in Fujian: achievements, problems and countermeasures. *Journal of Fujian Agriculture and Forestry University (Philosophy of Social Science)* 12 (4): 25–29.

Wang, S. L., S. X. Shi, R. Y. Wu, and Y. W. Dai. 2021. Rural ecology: an emerging discipline for the rural sustainable development(in Chinese). *Chinese Journal of Eco-Agriculture* 29 (12): 2116–2125.

Michael Warren.1991. Using indigenous knowledge in agricultural development. The World Bank, Washington, DC. https://documents1.worldbank.org/curated/en/408731468740976906/pdf/multi-page.pdf. Accessed September 18, 2024

Wezel, Alexander, Julia Goette, Elisabeth Lagneaux, Gloria Passuello, Erica Reisman, Christophe Rodier, and Grégoire Turpin. 2018. Agroecology in Europe: research, education, collective action networks, and alternative food systems. *Sustainability* 10:1214. https://doi.org/10.3390/su10041214.

Wiedenhoeft, M., S. Simmons, R. Salvador, et al. 2003. Agroecosystems analysis from the grass roots: a multidimensional experiential learning course. *Journal of Natural Resources and Life Sciences Education* 32:73–79.

Zurayk, Rami, Faraj el-Awar, Shady Hamadeh, Salma Talhouk, Christine Sayegh, Abdel-Ghani Chehab, and Kassem al Shab. 2001. Using indigenous knowledge in land use investigations: a participatory study in a semi-arid mountainous region of Lebanon. *Agriculture, Ecosystems and Environment* 86:247–262.

Chapter 9
Agroecology for Food Sovereignty: A Science, a Practice, and a Social Movement

Abstract The pursuit of providing the people of the world with available, accessible, nutritionally adequate, safe food is a technical problem. Food sovereignty is a different issue. Food sovereignty is a concept that emphasizes the right of people, communities, and countries to define their own agricultural, labour, fishing, food, and land policies that are ecologically, socially, economically, and culturally appropriate to their unique circumstances. It prioritizes local food production and consumption, ensuring that food systems are controlled by local producers rather than multinational corporations. Agroecology provides a systematic solution to food security, but this chapter explores the connection between agroecology and food sovereignty by defining agroecology as a science, a practice, and a social movement. It also revisits the original concept of agriculture from Chap. 1, where agriculture is defined as a unique life support industry, integrating politics into agroecology and providing an institutional guarantee for food sovereignty.

Keywords Agroecology · Agroecology for food sovereignty · Agroecology as a science · Agroecology as a set of practices · Agroecology as a social movement

Learning Objectives

After reading this chapter, student should be able to:

1. Describe the evolutionary history of the concept of agroecology.
2. Explain the linkages between agroecology and food sovereignty.
3. Describe the roles of agroecology as a science, a practice, and a social movement for developing food sovereignty.

****

> When farms are factories, they produce commodities and profit for agribusiness and charge external costs to the land and rural communities. When farms are natural habitats for humanity, domesticated crops and livestock, as well as wild plants and animals, they produce food and multiple other benefits to society.
>
> —Aldo Leopold (1966)

S. Wang, C. D. Caldwell, *Advanced Agroecology*,
https://doi.org/10.1007/978-3-032-08689-1_9

> To free ourselves, we must feed ourselves!
> —Anderson F (2018)

> Lutar! Construir Reforma Agrária Popular (in English: Struggle! Build People's Agrarian Reform).
> —MST's motto (Cited from Ricardo Serra Borsatto & Vanilde F. Souza-Esquerdo 2019)

Agroecology has a relatively short evolutionary history compared with other experimental sciences but currently is extremely relevant in the climate of an unsustainable global food system (Varghese 2022). The concept of agroecology started with applied ecological principles to crop sciences for better agricultural production in 1920–1970s (Benzin 1925; Klages 1928, 1942; Azzi 1928, 1942, 1956; Cox and Atkins 1979) and then evolved into the use of ecological principles to design and manage agroecosystems for balancing agricultural productivity and natural resource and environment maintenance in 1980–1990s (Altieri 1983, Altieri and Rosset 1995; Conway 1985; Carroll et al. 1990; Gliessman 1990, 1998). Entering the twenty-first century, the two milestone concepts of agroecology emerged, first in 2003, when agroecology was defined as ecology of entire food systems by extending agriculture to food systems (Francis et al. 2003), second in 2009, when agroecology was defined in hierarchical scales as a science, a practice, and a social movement (Wezel et al. 2009). Such multidimensional concepts of agroecology perfectly match the comprehensive concept of food sovereignty, which has been proposed by world farmers' organizations, led by La Vía Campesina (LVC) in 1996 (Lucantoni 2020; Gliessman 2020). This followed the comprehensive concept of food security provided by the FAO (1996); the LVC offered the formal definition of food sovereignty as (LVC 2007):

> The right of peoples to nutritious and culturally adequate food, accessible, produced in a sustainable and ecological way, and their right to decide their own food and productive system.

Ten year later, the world farmers' organizations, led by La Vía Campesina gathered in a village of Mali named Nyéléni, where the Nyéléni Declaration was presented (LVC 2007), stating that the fundamental prerequisites for attaining food sovereignty have three aspects: (1) Land reform and the defence of land and territory (political sovereignty); (2) The defence of national and local markets (economic sovereignty); and (3) Agroecology. That is to say, agroecology was expected to be the link between a state's economic and political sovereignty for ultimate food sovereignty. Defining agroecology as a science, a practice, and a social movement defines the role as shown in Fig. 9.1. This chapter addresses in detail the critical history and the link between agroecology and food sovereignty.

1 Agroecology as a Science

According to Wezel et al. (2009), since its inception in the 1920s, as an academic discipline within the realm of sciences, agroecology evolved from crop ecology, to agricultural ecology, to agri-food system ecology (i.e. the ecology of entire food

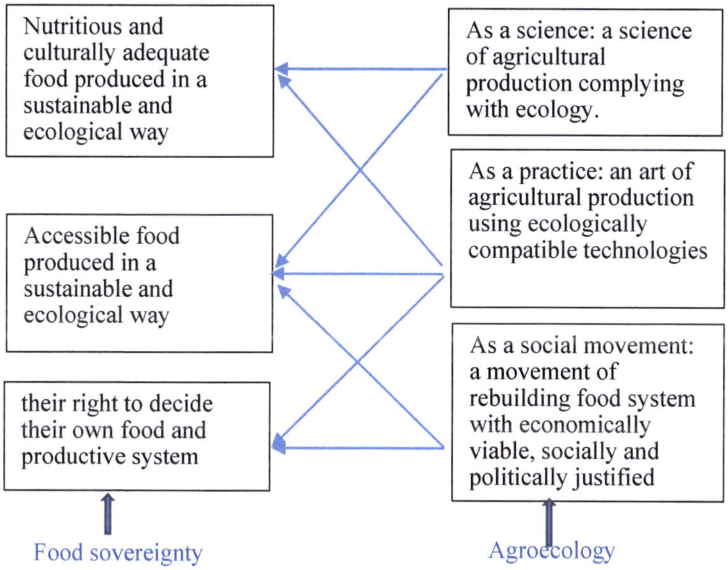

Fig. 9.1 How does the hierarchical concept of agroecology matches the multidimensional concept of food sovereignty?

system by Francis et al. (2003). Crop ecology dealt with crop production by using ecological methods; agricultural ecology designed and managed agroecosystem, and agri-food system ecology expanded ecology to design and manage food systems. The definition of entire food system ecology has been commonly recognized since it emerged (Stephen 2007). The latter can be characterized in the following aspects (Varghese 2022):

- Agroecology represents a comprehensive and integrative exploration of the ecology underlying the entire food system. This exploration takes into account multiple dimensions, including the ecological, economic, social, and political aspects, thereby providing a holistic understanding of the ecological dynamics of whole food systems globally, regionally, and locally.
- Agroecology involves the practical application of ecological concepts and principles. Specifically, these are utilized in the design and management of food systems that are sustainable, ensuring that they can meet present needs without compromising the ability of future generations to meet their own needs.
- Agroecology also emphasizes the integration of research, education, action, and transformative change. Through this integration, agroecology aims to foster sustainability across the ecological, economic, and social dimensions of food systems, creating a more balanced and resilient food production and consumption framework.

In contrast to various other conventional agricultural sciences, the ecological aspects of agroecology centre on the planning and design of farms and agroecosystems. This design is grounded in crucial ecological principles and is tailored to conform to the specific local environment. Altieri (2000) has described the six principles of agroecology:

- **Matter Recycling:** Improve the recycling of biomass, optimize the availability of nutrients, and strike a balance in the nutrient flow.
- **Soil Health**: Ensure that there are favourable soil conditions for the growth of plants, especially by effectively managing organic matter and boosting the biotic activity within the soil.
- **Resource Efficiency:** Reduce losses that occur as a result of the flows of solar radiation, air, and water. This can be achieved through micro-climate management, water collection techniques, and soil management strategies that involve increasing the soil cover.
- **Biodiversity Maintenance:** Implement species and genetic diversification within the agroecosystem both temporally and spatially.
- **Ecosystem Services:** Promote beneficial biological interactions and synergistic relationships among the components of agrobiodiversity, which in turn leads to the advancement of crucial ecological processes and services.
- **System Synergy**: Strengthen the 'immune system' of agricultural systems by enhancing functional biodiversity, including natural enemies, antagonists, and so on, through the creation of suitable habitats.

When the agroecology definition was updated from agroecosystem ecology to whole agri-food system ecology, the ecological principles guidelines were expanded correspondingly from the above 6 to 13 by the High Level Panel of Experts on Food Security and Nutrition (HLPE) in 2019 and grouped into 5 levels in accordance with FAO's Scaling Up Agroecology Initiative (SUAI) and 10 elements of agroecology for building a sustainable food system (FAO 2018a, b, c), as shown in Fig. 9.2.

The details are as follows:

1. **Recycling**: Research on resource circular utilization focuses on exploring effective methods to break the traditional linear resource consumption pattern and establish a closed-loop system that maximizes the value of resources throughout their life cycles.
2. **Input Reduction:** Study on minimizing or completely removing reliance on external inputs and enhance sufficiency of using those inputs if necessary.
3. **Soil Health:** Investigation into maintaining soil health and performance to facilitate better plant growth, with a focus on the manipulation of organic matter and the intensification of soil biological activity.
4. **Animal Health:** Knowing what, why, and how the animal has good life, and ensuring animal health physiologically and psychologically.
5. **Biodiversity:** Seeking the principles of preservation of biodiversity of species, functional diversity, and genetic resources and gaining the technologies to construct the overall biodiversity of agroecosystems across temporal and spatial dimensions.

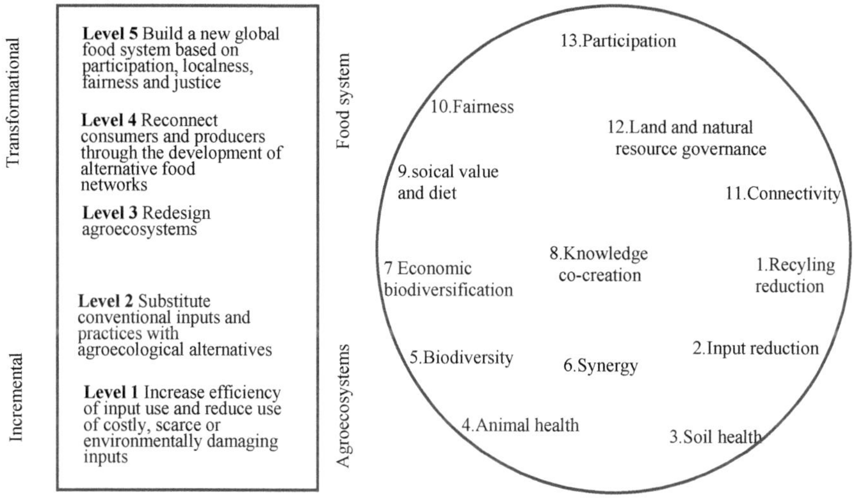

Fig. 9.2 Five levels of transition towards SFSs and related principles of agroecology (from HLPE 2019, URL: http://www.fao.org/3/ca5602en/ca5602en.pdf. Accessed May 25.2025, used with permission)

6. **Synergy:** Finding a way for strengthening the positive ecological interactions, synergies, integrative relationships, and complementary dynamics among the constituent elements of agroecosystems (i.e. animals, crops, trees, soil, and water).
7. **Economic Diversification:** Enhance the diversification of on-farm revenues through measures that foster increased financial autonomy for small-scale farmers and provide them with avenues for value-added activities. Simultaneously, create an enabling environment that allows these farmers to effectively meet consumer demands.
8. **Co-creation of Knowledge:** Experiencing the route for co-creation of knowledge in agroecology by means of participation and empowerment for stakeholders of agri-food system particularly farmers and females.
9. **Social Values and Diets:** Construct food systems that are grounded in the culture, identity, traditions, as well as social and gender equity of local communities with the provision of healthy, diverse, seasonal, and culturally suitable diets.
10. **Fairness:** Ensure that all participants in food systems, with a particular emphasis on small-scale food producers, are able to maintain dignified and sustainable livelihoods. This should be achieved through the implementation of fair-trade practices, the establishment of equitable employment conditions, and the just treatment of intellectual property rights.
11. **Connectivity:** Foster closeness and trust between producers and consumers by advocating fair and streamlined distribution networks and reintegrating food systems into local economic frameworks.
12. **Land and Natural Resource Governance**: Finding the institutional mechanisms for advancement, encompassing the acknowledgment and assistance of

family farmers, small-scale agricultural producers, and peasant food producers who serve as stewards of natural and genetic resources in a sustainable manner.
13. **Participation:** Foster the engagement of multifaceted social organizations and amplify the deliberative participation of food producers and consumers in decision-making frameworks.

Among the abovementioned 13 principles of agroecology as a scientific discipline, participation (principle#13) and co-creation of knowledge (principle#8) interact and are significant aspects for sustainable food system development. They flow from an understanding and appreciation for indigenous farming systems that adhere to some or all of these fundamental principles. Most indigenous communities find themselves constrained from sustaining these traditional practices. Policies concerning food, agricultural activities, seeds, water resources, and the use of agrichemical inputs have created incentives that have forced a departure from these holistic food system practices (CIDSE 2018). Instead, there has been a move towards practices that place exclusive emphasis on enhancing the productivity of a limited number of chosen crops or livestock (see Chap. 8).

Box 9.1 depicts that agroecology as a science implies to technology innovation for agri-food system transformation.

Box 9.1 Exploring Barriers to the Agroecological Transition in Nicaragua: A Technological Innovation Systems Approach (Schiller et al. 2019)

- **Background**
  Agroecological farming systems, which apply ecological principles for sustainable farm design and management, are seen as promising alternatives to input-intensive agro-industrial models. In Nicaragua, an agroecological transition is underway. The agroecological innovation system (AeIS) approach is introduced to analyse the systemic barriers and interaction-caused blockages in Nicaragua's agroecological transition.
- **Processes**
  Theoretical Framework: The agroecological innovation system (AeIS) is a subset of the national agricultural innovation system. A structural-functional analysis of the AeIS is conducted, which distinguishes between structures (actors, institutions, interactions, and infrastructures) and functions (such as experimentation by entrepreneurs, knowledge development, etc.). Systemic problems are related to the presence/absence and quality/capacity of structures in contributing to functions, and these problems can form blocking mechanisms and vicious cycles.
  Methodology: A qualitative case-study analysis is carried out. Data is collected through field work in northern and Central Nicaragua from 2014 to 2018, including semi-structured interviews, workshops, farm visits, and a

(continued)

Box 9.1 (continued)

literature review. The data is analysed using event history analysis to explore the development of the AeIS over time and to understand the performance of the seven functions of the technological innovation systems.
Analysis of the AeIS Development: The growth of the AeIS in Nicaragua is divided into three phases influenced by the ideology of the ruling party. Since 2006, the agroecological transition has gained momentum, with more favourable policies and the growth of relevant platforms and networks. However, a structural-functional analysis of the seven functions of the AeIS shows that there are many problems. For example, in experimentation by entrepreneurs, farmers face difficulties such as lack of market opportunities and infrastructure. In knowledge development, there is a heavy dependence on donor funding. In market formation, input markets develop slowly, and output markets have issues like low consumer purchasing power and lack of product differentiation.

- **Achievements**
  Identifying Barriers and Cycles: The research identifies the main barriers to the agroecological transition in Nicaragua, including weak guidance in the search for agroecology, insufficient resources in terms of capacity and quantity, and lacking market development. It also reveals how these barriers interact to form vicious cycles that hinder the further diffusion of agroecology. For example, the lack of a common vision for agroecology negatively impacts all other functions in the system, and hard and soft institutional problems affect entrepreneurial experimentation and market formation.
  Contributions to Policy and Practice: The study provides suggestions for promoting agroecology in Nicaragua. It emphasizes the need to create a 'thick legitimacy' for agroecology, strengthen policy coordination, involve non-traditional actors in agroecological development, address farmers' practical problems (such as land tenure and access to resources), and improve consumer awareness of agroecological products. These suggestions can guide future policy-making and practical actions to support the agroecological transition in Nicaragua.

In summary, regarding agroecology as a science, four decades of empirical scientific research have firmly established that agroecological farming systems, particularly those encompassing diversified agricultural practices, represent the most efficacious approach within the realm of agriculture to address the multifaceted environmental challenges imperilling global food security in the future. These challenges include climate change, which disrupts weather patterns and temperature regimes; soil erosion, which degrades soil quality and fertility; water scarcity, which limits the availability of this crucial resource for irrigation and other agricultural needs; and the loss of biodiversity, which undermines the resilience and stability of ecosystems upon which agriculture depends (Varghese 2022).

2 Agroecology as a Practice

When Wezel et al. (2009) first defined agroecology as a practice, they just referred to the technologies stemming from ecological principles for seeking environmentally friendly or sustainable agriculture. Their focus was on (Varghese 2022):

- Agroecology, as an array of agricultural practices, endeavours to identify approaches for enhancing agricultural production, achieved by capitalizing on the inherent ecological processes and the functions of ecosystems, in order to diminish the fossil energy and its product like fertilizer input.
- Agroecology optimizes functional biodiversity and fortifies the biological regulatory mechanisms within agroecosystems. It strives to reduce the use of synthetic and harmful insecticide, as well as to minimize the generation of waste.
- Agroecology integrates newly developed technology and traditional or indigenous knowledge to generate the adaptive technology to climate change and market fluctuation.

These practices that stem from ecology and attempt to create a new agricultural model have had various names but are collectively known as alternative agricultures (Vamdemeer 1995; Altieri 1987). These include organic agriculture (originated in the USA and some European countries), bio-dynamic agriculture (originated in Germany), nature farming (originated in Japan), regenerative agriculture (originated in the USA), sustainable intensification agriculture (originated in some European countries like England), and ecological agriculture (China and some east European countries). The turning point for unification occurred in 2011 when the UN Special Rapporteur on the Right to Food, Olivier de Shutter submitted the report named ‘agroecology and right to food’ to the UN human rights committee, in which he endowed agroecology as a practice being synonymous with ecological agriculture as a unified model under the umbrella of agroecology as a science (de Shutter 2011a, b, 2012, 2014). According to Varghese (2022), agroecology as practice or eco-agriculture is quite different from other kinds of alternative agricultures:

- **Agroecology as a Practice Differs from Organic Agriculture**: Agroecology is often confused with the very important initiatives of organic agriculture or regenerative agriculture. It is important to understand the distinction between agroecology and the other approaches to sustainability. First, let’s compare agroecology with organic agriculture. Agroecology considers farmland as an ecosystem, accounting for soil health, water quality, air quality, biodiversity, and more. The goal is to create stable, resilient food production systems and to treat farmland as complex ecological webs. Organic agriculture primarily focuses on producing food without synthetic chemicals and follows specific standards for organic practices. It has come under some criticism because many large-scale organic farms do not manage ecosystems holistically but rather by prescription. Certainly, the food from ecologically minded growers can align with agroecological principles.

- **Agroecology as a Practice Differs from Regenerative Agriculture:** Concerning the comparison of agroecology and regenerative agriculture, agroecology is both a field of scientific research and the set of farming practices. It emphasizes sustainable and ecologically sound farming and integrates ecological principles into agricultural systems. Therefore, it embraces social, cultural, and economic aspects of agriculture. At present, it is widely used in developing countries rather than developed countries. The objective of agroecology is to enhance biodiversity, promote social justice, and improve food systems. Regenerative agriculture prioritizes soil health and enrichment. It is a holistic farming approach focused on soil regeneration, biodiversity restoration, and ecosystem health improvement. Its major principles include restoring, conserving, and enhancing natural resources, minimizing waste, and creating self-sustaining cycles. Production techniques commonly include cover cropping, no-till farming, organic production, and managed grazing. Therefore, the goal of regenerative agriculture is not just food production but also improving ecosystem services.
- **Agroecology as a Practice Differs from Sustainable Intensification agriculture:** Sustainable intensification (SI) places its emphasis on enhancing the productivity per unit area of land. At the same time, it endeavours to take into consideration certain environmental externalities that stem from the present model of industrial agriculture. This approach represents a prevalent and frequently adopted stance by corporations in response to the ecological crisis within food systems. Its primary orientation remains centred on productivity, with the aim of increasing the output on an individual plant or animal basis. Additionally, it employs technological interventions to boost the efficiency of resource utilization for specific inputs, such as water, labour, or agrochemicals. However, SI lacks a holistic perspective. It fails to adequately consider the potential adverse effects that technical interventions might have on the farming system, including the ecological equilibrium of agroecosystems. For instance, in the context of SI practices, no-till farming is commonly associated with the application of herbicides. While no-till farming can contribute to carbon sequestration, particularly when integrated with other practices like the cultivation of cover crops, and the use of herbicides can enhance the labour-use efficiency on the farm, the overall ecological impact of no-till farming in this scenario is detrimental. This is because the use of herbicides can have far-reaching consequences, affecting biodiversity, water quality, the ecological balance within agroecosystems, and even posing risks to human health. Most significantly, those who engage in SI do not question or challenge the concerning issue of corporate concentration within food systems. Whereas agroecology places emphasis on minimizing inputs and promoting diversity, along with social and political change. Agroecology is centred on enhancing both ecological and human well-being and addressing concerns regarding equity and governance.

Overall, agroecology (eco-agriculture), organic agriculture, regenerative agriculture, and sustainable intensification agriculture were developed as practices to contribute to better agri-food systems and finally to a healthier planet. They differ in the specific focus and methodologies; agroecology encompasses broader dimensions than the other three.

Despite barriers, there are several successful examples of integrating agroecology into mainstream agriculture around the world. Here are a few notable ones (abstracted from Sinclair et al. 2019 with exception of Zimbabwe and China's cases):

- Andhra Pradesh, India: The state government has implemented a large-scale agroecology program known as Zero Budget Natural Farming (ZBNF). This initiative promotes chemical-free farming by using natural inputs and has reached millions of farmers, significantly improving soil health and reducing input costs.
- Sahel Region, Africa: Farmer-Managed Natural Regeneration (FMNR) has been widely adopted to combat desertification. This practice involves protecting and managing the natural regrowth of trees and shrubs, which has led to the regreening of vast areas, improved soil fertility, and increased crop yields.
- Peru: The government has supported agroecological practices through policies that promote land restoration and biodiversity conservation. These efforts have helped to enhance food security and resilience to climate change.
- France and Switzerland: Both countries have implemented policies to support agroecological practices, focusing on sustainable land management and biodiversity conservation. These policies have encouraged farmers to adopt practices that enhance ecosystem services and reduce environmental impacts.
- Zimbabwe: Zephaniah Phiri, known as the 'Water Harvester', transformed a dry parcel of land into a fertile oasis using agroecological methods, such as water management, intercropping, and crop rotation. His success has inspired many local and international farmers (https://futureoffood.org/insights/agroecology-case-studies/).
- Mainland China: China started ecological agriculture (or agroecology as a practice) since the reform and opening time in the early 1980s via the top-down route, even though agroecology practice or ecological agriculture was hampered with the core development of the economy, it moved slowly with more than 10 eco-agriculture models nationwide. It was after the eco-civilization strategy was set up as a national strategy and the awakening consumer awareness food safety, that China' eco-agriculture was rejuvenated in both top-down and bottom-up routes (see Chap. 10).

These examples demonstrate the potential of agroecology to create sustainable and resilient agricultural systems. They are very concrete examples of how we can move forward with the adoption of agroecology around the world.

Integrating agroecology into mainstream agriculture presents several challenges (Wezel et al. 2018):

- Lack of Awareness and Understanding: Many farmers and stakeholders are not fully aware of agroecological practices or their benefits.
- Economic Barriers: Transitioning to agroecological methods can involve initial costs and potential short-term yield reductions, which can be a deterrent for farmers.

- Policy and Institutional Support: There is often a lack of coherent policies and institutional support to promote and sustain agroecological practices.
- Knowledge and Training: Effective implementation requires extensive knowledge and training, which are not always readily available.
- Research and Development: There is a need for more research to develop and refine agroecological practices and to demonstrate their long-term benefits.
- Market Access: Farmers may face challenges in accessing markets that value and pay a premium for agroecologically produced goods.
- Cultural and Social Factors: Traditional farming communities may be resistant to change, and there can be social and cultural barriers to adopting new practices.

The adoption of agroecological principles and practices into mainstream agriculture will integrate sustainable practices that enhance biodiversity, soil health, and ecosystem services into conventional farming systems. Therefore, there are some strategies to be considered (Wezel et al. 2020; Niggli et al. 2021):

- Promote Biodiversity: Encourage crop diversification and the use of cover crops to improve soil health and reduce pest pressures.
- Reduce Chemical Inputs: Minimize the use of synthetic fertilizers and pesticides by adopting organic farming practices and using natural pest control methods.
- Enhance Soil Health: Implement practices like crop rotation, composting, and reduced tillage to maintain and improve soil fertility.
- Foster Knowledge Sharing: Facilitate the exchange of knowledge between farmers, researchers, and policymakers to promote the adoption of agroecological practices.
- Support Policy Changes: Advocate for policies that support sustainable farming practices, such as subsidies for organic farming and research funding for agroecological methods.
- Encourage Community Involvement: Engage local communities in the planning and implementation of agroecological projects to ensure they meet local needs and conditions.

3 Agroecology as a Social Movement

Agroecology, as a dynamic movement, endeavours to revolutionize the agricultural landscape. Its overarching objective is to construct food systems that are not only locally appropriate but also resilient and sustainable. These systems aim to enhance the economic viability of rural regions by relying on short marketing chains, ensuring equity, and promoting both fair and secure food production processes.

At its core, this movement places significant emphasis on supporting the diverse manifestations of smallholder food production and family farming. This encompasses a wide range of individuals, such as farmers, pastoralists, fishers, and gatherers, all of whom play crucial roles in the food production ecosystem.

The specific goals of the agroecology movement are multifaceted. It advocates for food sovereignty, which empowers communities to have control over their food systems. It also emphasizes the importance of respecting and safeguarding local knowledge, which has been accumulated over generations and is integral to sustainable agriculture. Social justice is another key tenet, ensuring that all stakeholders in the food chain are treated fairly. Additionally, the movement seeks to preserve local identity and culture, which are often intertwined with traditional agricultural practices. Finally, it upholds the rights to local and Indigenous seeds and breeds, recognizing their unique genetic diversity and their significance in maintaining sustainable and resilient food systems.

There two quite successful cases for agroecology movement are listed in Boxes 9.2 and 9.3.

Box 9.2 Agroecological Movement Case#1

The Campesino-to-Campesino agroecology movement of ANAP in Cuba (Rice and Einbinder 2023; Lucantoni 2020; Rosset et al. 2011).

- **Background**
  In the context of the collapse of the socialist bloc in Europe and the tightening of the US trade embargo, Cuba faced a severe agricultural crisis. Its traditional agricultural model, which was highly dependent on imported inputs and monoculture exports, became unsustainable. Peasant organizations worldwide, especially those in **La Via Campesina**, were increasingly interested in agroecology as an alternative, which emphasizes ecological principles, local knowledge, and reduced external inputs, and offered a potential solution for Cuba to boost food production and achieve food sovereignty. Such agroecology movement was led by a farmer organization in Cuban, The National Association of Small Farmers (ANAP).
- **Process**
  ANAP in Cuba learned about the Campesino-a-Campesino (CAC, peasant-peasant organization) methodology from Nicaragua in the mid-1990s. They started a trial project in Villa Clara province in 1997 with external donor funding. Promoters, who were farmers recognized for their successful agroecological practices, and facilitators, responsible for logistical arrangements, played key roles in the initial phase. In 2001, ANAP decided to transform CAC from a project into a national movement, aiming to mobilize peasant energy and creativity. This led to the spread of the movement across all Cuban provinces. As it grew, new roles like coordinators were created, and a five-step process was formalized. The movement also classified farms based on their agroecological integration level and developed innovative methods like the 'Banes Method' to enhance its effectiveness. For example, The 'El Palmar' farm, located in Pinar del Río, Cuba,

(continued)

Box 9.2 (continued)

started its agroecological transition. It diversified agricultural production by planting various fruit trees, local crops, and medicinal plants. The family also adjusted animal production, with a focus on sustainable feeding. For pest and disease management, they employed cultural, biological, and some chemical control methods (exclusively for tobacco). In soil management, strategies like crop rotation, composting, and vermiculture were used to improve soil quality. The farm's water supply relied on nearby water sources, and they aimed to enhance water-use efficiency. According to the ANAP classification, the farm was in Category 2, meaning it was in the process of transition.

- **Achievements**

The ANAP agroecology movement achieved remarkable results. In terms of growth, the number of families joining the movement increased from over 200 in 1999 to 110,000 in 2009, accounting for about one-third of peasant families in Cuba. The productivity of agroecological farms and the peasant sector in general significantly improved. Data from Sancti Spíritus province showed that farms with a higher level of agroecological integration had greater production value per worker and per hectare. The peasant sector's overall production also increased, contributing more to national food production. Moreover, agroecological farms demonstrated greater resilience to climate change. After Hurricane Ike in 2008, more agroecological farms suffered less damage and recovered faster. Additionally, the movement had an impact on the peasant family structure, diversifying roles and reducing patriarchy. However, the movement also faced challenges such as achieving gender equality and avoiding over-bureaucratization. Meanwhile, the transition to agroecology brought multiple positive results. In terms of food security, it improved the availability, access, stability, and utilization of food. The family's diet became more diverse, and they became less dependent on the market. The soil health of the farm improved, with increases in organic matter content and cationic exchange capacity, though some elements still needed improvement. Environmentally, it enhanced sustainability by reducing greenhouse gas emissions and improving biodiversity. Economically, although the initial transition had high costs, in the medium to long term, it was expected to increase the family's net income. Overall, the farm's transition to agroecology led to better living conditions for the Rodríguez family and was a step towards achieving food sovereignty in Cuba. The results also showed diverse livelihood outcomes. Many farmers maintained or increased production by focusing on local subsistence crops and animal products, and most had sufficient household consumption, mainly due to increased subsistence production and revaluing traditional food practices. Although some farmers had no sales during the pandemic, others continued selling through alternative

(continued)

Box 9.2 (continued)

strategies. Statistical analysis indicated that prior engagement with an organization was positively correlated with agroecological practices (such as reduced agrochemical use, increased crop diversification, and higher subsistence production), as well as some livelihood outcomes (increased production level and sufficient consumption). Qualitative data revealed that agroecological practices and social networks within organizations helped farmers access informal markets, pivot to new products, and engage in bartering. In simpler terms, for Cuba, a nation that still remains significantly reliant on importing food from the global market, the shift towards agroecology by the vast majority of its family farms represents the most rational and practical approach to attaining national food sovereignty, as this transition towards agroecology progresses, an increasing number of farming families are becoming part of the 'Agroecological revolution'. The overarching national goal is to bring 1.5 million hectares of land under agroecological management. Achieving this target would be adequate to secure Cuba's food sovereignty.

Box 9.3 Agroecological Movement Case#2

Brazil's Landless Rural Workers' Movement (MST) (Lagier 2020; Borsatto and Souza-Esquerdo 2019; Alaniz and Dal Ri 2019).

- **Background**
  The Brazilian Landless Rural Workers' Movement (MST), established in 1984, a significant part of the global food sovereignty coalition La Via Campesina (LVC), plays a crucial role in the agricultural and social landscape of Brazil. Agroecology, which has multiple dimensions, is an important aspect within the MST. Agroecology provides theoretical and practical support for constructing a new agricultural and food system in Brazil through the MST series activities.
- **Process**
  The MST began to explore new development models in the 1990s and started promoting agroecology in the mid-1990s by viewing agroecology as a way to build a more just agricultural system. Also, the MST has been actively involved in a range of educational initiatives, aiming to safeguard the right to education for rural children, youth, and adults, while also providing the necessary technical and political training to support the movement's socioeconomic development and strengthen its social influence. In 2000, the MST, during its IV National Congress, decided to promote agroecology as a strategic approach to rural development. Subsequently, the Milton Santos School in Paraná was founded in partnership with the

(continued)

Box 9.3 (continued)

Federal University of Paraná. It started offering a post-secondary technical course in agroecology in 2003, with the construction and operation of the school relying on the collective work of students, educators, and MST-affiliated workers. Since then, the MST has carried out various activities to promote agroecology, such as organizing events, establishing schools for agroecological training, and forming partnerships with different organizations. However, in the process of promoting agroecology, it faces challenges in many aspects. For example, some government support policies for family farming make farmers less aware of the crisis, which restricts the promotion of agroecology; the large-scale, heterogeneous, and geographically dispersed characteristics of the MST make it difficult to spread agroecological practices; there are conflicts in cooperation with external allies, and government policies in support of agroecology are limited.

From May 2017 to April 2018, the researcher conducted fieldwork at an MST-affiliated land reform settlement in Parana. This settlement is well-known for its agroecological production and is home to the Ecological Resistance School (ERS), a transnational activist school for agroecology. The research methods included 84 in-depth semi-directed interviews and participant observation. Informed consent was obtained from all individual participants, and the name 'Ecological Resistance School' is a pseudonym.

The MST's concept of professional education stands in contrast to the official model. It emphasizes the connection between teaching and real productive work, integrating technical and political training with the needs of collective development and the struggle for agrarian reform. At the Milton Santos School, the education process is centred around cooperation and democratic management. The school's organizational structure, similar to that of the MST, is based on horizontal communication and decision-making, with various collectives and working groups involved in different aspects of school management. The education-work link is a fundamental principle in the MST's pedagogy. Students are engaged in real-world work, both in maintaining the school and in practical activities in rural settlements through the 'pedagogy of alternation'. This approach involves alternating between periods of study at school and practical work in the community, where students apply their learned knowledge and skills.

- **Achievements**

  The MST has achieved certain progress in promoting agroecology, such as the multiplication of agroecological production experiences in some regions, The MST has strengthen knowledge exchange among peasants, increased the construction of demonstration areas, and promoted cooperation with urban sectors and the development of local markets. The experience of the MST in promoting agroecology provides valuable lessons for other regions and has become a source of social polarization. Young LVC

(continued)

Box 9.3 (continued)

activists in an agroecology bachelor's program faced complex internal social obstacles, which severely undermined their ability to be effective agents of socioecological change. Meanwhile, the technical training in agroecology at the Milton Santos school has proven to be successful. Most graduates return to their communities and apply agroecological science in production, not as salaried workers but as active members of the community. The school's operation, based on the principles of cooperation and democracy, has effectively deconstructed the concept of alienated work, inspiring collective forms of management and social organization. It has also contributed to the advancement of the struggle for agrarian reform, modifying rural production processes and labour relations. Overall, the school represents a successful example of agroecology-based education within the new productive matrix defined by the MST, making a significant contribution to rural vocational education.

4 Agroecology for Food Sovereignty: Academic and Policy Implications

The Nyéléni Declaration (LVC 2007, cited from Anderson F. 2018) proposed six pillars of food sovereignty to broaden food security concept by FAO (1996), there are:

- **Food as Human Right:** Food sovereignty elevates the entitlement of every individual, community, and population group, such as the hungry, those living under occupation, in conflict-affected regions, and marginalized communities, to access adequate, nutritious, and culturally suitable food. This promise is enshrined as the cornerstone of policies governing food production, agriculture, livestock rearing, and fisheries. It firmly opposes the notion that food should be regarded merely as a commodity or an element within the international agribusiness framework.
- **Valuing Food Providers:** Food sovereignty acknowledges and upholds the significant contributions made by diverse groups of people involved in the food production and processing chain. It recognizes and respects the rights of both women and men, encompassing peasants, small-scale family farmers, pastoralists, artisanal fisherfolk, forest dwellers, indigenous peoples, as well as agricultural and fisheries workers, including migrants. These individuals play crucial roles in cultivating, growing, harvesting, and processing food. Conversely, food sovereignty opposes policies, actions, and programs that devalue the efforts of these groups, pose threats to their livelihoods, and aim to marginalize or eliminate them from the food-related sectors.
- **Place-Based Food Systems**: Food sovereignty fosters a stronger connection between food providers and consumers, effectively bridging the gap between

these two key stakeholders in the food system. It places food providers and consumers at the very heart of the decision-making processes concerning food-related matters. This concept serves as a safeguard for food providers, shielding them from the detrimental impacts of food dumping and excessive food aid in local markets, which can undermine their livelihoods and the viability of local food production. For consumers, food sovereignty acts as a protective measure against substandard and unhealthy food products. It also guards them against inappropriate food aid that may not meet their actual nutritional needs or cultural preferences, as well as food that has been contaminated or genetically modified in ways that could pose risks to their health and well-being. Furthermore, food sovereignty stands in opposition to governance structures, agreements, and practices that are founded upon and perpetuate unsustainable and unfair international trade models. These existing systems often empower distant and unaccountable corporations at the expense of local food producers and consumers, and food sovereignty resists such imbalances of power, advocating for a more equitable and sustainable food system.

- **Food Autonomy:** Food sovereignty empowers local food providers by entrusting them with the management and control of various crucial resources, including territory, land, grazing areas, water sources, seeds, livestock, and fish populations, while also upholding their rights. These providers are enabled to utilize and share these resources in manners that are both socially equitable and environmentally sustainable, with a particular emphasis on preserving biodiversity. It acknowledges the reality that local territories frequently transcend geopolitical boundaries and guarantees the right of local communities to reside in and make use of their respective territories. Moreover, food sovereignty fosters constructive interactions among food providers from diverse regions, territories, and sectors. Such interactions play a vital role in facilitating the resolution of internal disputes, as well as conflicts that may arise with local and national authorities. Additionally, food sovereignty firmly opposes the privatization of natural resources, which can be achieved through the implementation of laws, the establishment of commercial contracts, and the enforcement of intellectual property rights regimes.
- **Building Food Knowledge and Skill:** Food sovereignty is founded upon the expertise and indigenous knowledge possessed by food producers, as well as the capabilities of their local institutions. These entities play a crucial role in preserving, advancing, and overseeing localized food production and harvesting methodologies. They are also instrumental in establishing relevant research frameworks to bolster these practices and in ensuring the transmission of this accumulated wisdom to subsequent generations. Additionally, food sovereignty firmly opposes technologies that have the potential to undermine, pose a threat to, or contaminate these traditional systems. For instance, genetic engineering is among the technologies that are deemed incompatible with the principles of food sovereignty due to its potential negative impacts on local food production and harvesting systems.

- **Connecting Food with Nature:** Food sovereignty leverages the offerings of nature through a variety of low external input agroecological production and harvesting techniques. These approaches aim to optimize the role played by ecosystems, thereby enhancing resilience and the capacity for adaptation, particularly in the context of climate change. It endeavours to restore the health of the planet, recognizing that a healthy planet is essential for our own well-being. Moreover, food sovereignty explicitly opposes methods that undermine the beneficial functions of ecosystems. This includes practices that rely on energy-intensive monocultures and livestock factory systems, as well as destructive fishing practices and other industrialized production methods. Such practices not only cause environmental harm but also contribute significantly to the acceleration of global warming.

In summary, agroecology delves into the origin of food, the methodologies of its production, the individuals engaged in production, and the mechanisms through which knowledge and skills pertaining to food and agriculture are disseminated. This field of study contextualizes the comprehension of food and agricultural production within ecological, social, cultural, and political frameworks. By doing so, it elucidates the diverse manifestations of agriculture and acknowledges the significance of this diversity.

Scientifically and philosophically, the comprehensive definition of agroecology as a science, a practice, and a social movement is intricately interconnected. The agroecology practice and movement rely on its scientific knowledge and principles to drive the promotion and implementation of an agricultural methodology. Owing to the reciprocal connections and co-evolutionary processes among these three aspects, agroecology represents a comprehensive and integrated approach. This approach is centred around the transformation of agri-food systems, aiming to foster more equitable and sustainable agri-food systems. Moreover, the advancement of agroecological practices is also achieved through the collaborative co-creation of knowledge, involving scientists, farmers, government officer, and NGO members. By bringing together different stakeholders and expertise, this collaborative process enriches the understanding and implementation of agroecology, leading to more effective and sustainable outcomes for agri-food systems.

Socially and politically, agroecology endeavours to establish robust food democracies, grounded firmly in local, experiential, and Indigenous knowledge systems. These knowledge systems are deeply rooted in specific places and are often passed down through generations, embodying the wisdom of traditional communities. The practical applications of agroecology are strengthened through intercultural interactions that are characterized by mutual respect. Such interactions occur among food producers, individuals working within the food system, and consumers. Special consideration is given to women, recognizing their significant contributions and unique roles in food production and consumption.

The abovementioned multifunction of agroecology can be narrated by its biological, economic, social, and political dimensions of agroecology.

4.1 Agroecology Guides the Establishment of Low-Carbon Food Production System

As a science, agroecology is based mainly in biology and agronomy. It offers an integrative perspective on humanity's role within natural cycles and elucidates the imperative for farming systems to strengthen the ecosystems and societies upon which they rely. Therefore agroecology plays a crucial role in constructing a low-carbon agricultural production system, reflecting the following aspects. Firstly, agroecology promotes biodiversity, which is fundamental to a low-carbon agricultural system. By diversifying species and genetic resources in agroecosystems over time and space, it enhances beneficial biological interactions. For example, intercropping systems, a common practice in agroecology, can improve resource use efficiency (Nicholls et al. 2016). Different crops with varied root systems and growth patterns can better utilize nutrients, water, and sunlight, reducing the need for excessive external inputs, such as fertilizers and pesticides. These inputs often require significant energy for production and application, and their overuse leads to excessive greenhouse gas emissions. Thus, biodiversity—enhanced by agroecology—helps in reducing the carbon footprint of agriculture. Secondly, agroecology emphasizes the recycling of biomass and optimization of nutrient cycling. Practices like compost application and cover cropping are integral agroecological practices. Compost, rich in organic matter, not only provides essential nutrients for plants but also improves soil structure and water-holding capacity. When applied to the soil, it reduces the need for synthetic fertilizers, which are energy-intensive and can release nitrous oxide, a potent greenhouse gas. Cover crops, however, can fix nitrogen in the soil, prevent soil erosion, and increase soil organic matter content. This efficient nutrient cycling system minimizes the loss of nutrients and reduces the reliance on external, carbon-intensive inputs, thereby contributing to a low-carbon agricultural production system. Thirdly, agroecology contributes to carbon sequestration in the soil. Soils managed under agroecological principles tend to have higher organic matter content. As plants grow, they absorb carbon dioxide from the atmosphere through photosynthesis. A portion of this carbon is then transferred to the soil through root exudates and plant residues. The complex food webs in agroecological soils, populated by beneficial microorganisms, help to decompose and store this organic matter for long periods. This process of carbon sequestration in the soil not only mitigates climate change by removing carbon dioxide from the atmosphere but also improves soil fertility and resilience (Wang et al. 2010). In this way, agroecology actively participates in building a low-carbon agricultural production system that is both environmentally friendly and sustainable.

4.2 Agroecology Reverses Economic Discrimination in the Agri-Food Value Chain

Agroecology plays a crucial role in building sustainable food systems, and its economic impacts are multi-faceted. Firstly, in terms of economic principles, agroecology emphasizes financial independence. Farmers and agricultural organizations strive to be in control of economic and technical decisions. This promise helps reduce the influence of external economic forces, ensuring the stability of the agricultural economy. Another important principle is market access and autonomy. Practices allied to this principle enable agricultural entities to adapt to market changes and maintain economic viability. Secondly, agroecology contributes to sustainable food systems by enhancing the overall economic resilience. The principles of sustainability and adaptability in agroecology are achieved through the diversification of production systems. This knowledge-based approach improves the efficiency of resource utilization and economic returns, making the food system more sustainable in the long run. Finally, agroecology promotes economic equity within the food system. The principle of social equity is reflected in various ways, including an emphasis on local production and consumption. By achieving economic equity, agroecology practices help to build a more inclusive and sustainable food system that can meet the needs of all stakeholders.

In conclusion, agroecology's economic principles, including financial independence, market access, and autonomy, contribute to the resilience and equity of the food system, making it an essential approach for building sustainable food systems. Among them, equitable food trade systems are the key to reach the goal, which is worth addressing.

For a considerable period, international peasant organizations, like the Via Campesina, have advocated for the exclusion of agriculture from all trade agreements. Nevertheless, such an approach might create a default situation that is unjust to the diverse stakeholders within the food system. This is because individuals and nations rely on one another when it comes to obtaining food, highlighting the interdependence that exists in the food access context. Considering that agroecological transformations of our food systems are essential, trade regulations should be crafted in a way that safeguards agroecological food systems from being overwhelmed by large volumes of cheap imports that are produced in an unsustainable manner. Additionally, these trade rules should encourage (rather than impede) collaborative efforts among different parties. This would entail reforming trade agreements so that they place greater emphasis on public policies. These policies should give precedence to the advancement of agroecology, climate mitigation measures, and the protection of human rights, over the mere commercial movement of goods, services, and investment. Therefore, one of the basic tenets of agroecology is to move control of the food system from a global approach to a local approach. The intent is to make value chains fairer and more sustainable. To move in that direction, there are several key strategies (Dumont et al. 2016, 2021):

- **Adopting fair-trade practices** will reduce poverty and improve living standards; value chains need to ensure that producers, especially small holder farmers, receive fair prices for their products.
- **The predominant linear 'take-make-dispose' model** for value chains is obviously non-sustainable both from an environmental and economic standpoint. There is a need to adopt a circular economy principle which provides a 'take-make-use-recover' model. This involves recycling materials, reducing waste, and designing products for longevity.
- **Social, ethical, and environmental standards** are essential. This requires efficient certifications, audits, and partnerships among the stakeholders in the value chains. However, the necessity for transparency and traceability must not impose a bureaucratic burden on stakeholders because it can produce conflict and delays within the value chain. It will require cooperation.
- **Supporting Local Economies:** Utilizing local materials and labour is good for the environment by reducing carbon footprint but also is good for the economy and social fabric by supporting local communities.
- **Integration of Related Value Chains and Webs:** Various stakeholders, governments, NGOs, and businesses access are key to a number of value chains. There is a need for collaboration across sectors.

4.3 Social Dimension of Agroecology Linked to Sustainable Development Goals

Agroecology is deeply rooted in the culture, identity, tradition, innovation, and knowledge of local communities. It is knowledge-intensive and promotes horizontal farmer-to-farmer contacts for sharing knowledge, skills, and innovations. It also encourages alliances between farmers and researchers on an equal footing. Agroecology creates opportunities for solidarity and discussion among culturally diverse peoples, including different ethnic groups, as well as between rural and urban populations. It respects diversity in terms of gender, race, sexual orientation, and religion. Specifically, it offers opportunities for young people and women and encourages women's leadership and gender equality. Moreover, agroecology does not necessarily rely on expensive external certification. Instead, it often depends on producer-consumer relations based on trust and promotes alternative models like PGS (Participatory Guarantee System) and CSA (Community Supported Agriculture). It also supports communities in maintaining their spiritual and material connection with the land and environment.

Therefore, starting from the existing knowledge and skills of farmers and food producers, agroecology is well-suited to fulfilling their right to food. It can develop appropriate technologies for small-scale farmers, peasants, indigenous people, and other similar communities. In developing countries, as agriculture is a common occupation, agroecology can reverse rural-to urban migration and family fragmentation. By involving people in agroecological practices and value chain development, it can make rural life and food production more attractive, contributing to local economic development, social cohesion, and stability.

Placing food producers at the centre of food systems, agroecology gives new value to farmer identities and strengthens their confidence and involvement. It also restores justice to the food system by bringing producers and consumers closer in shorter value chains, reducing the influence of corporate power (Anderson 2018). This promotes trust and solidarity in the producer-consumer relationship, provides nutritious and culturally-appropriate food, and protects local cultural identities. Additionally, agroecology offers women opportunities to increase economic autonomy and influence power dynamics within the family. However, since its impact on gender relations is not always positive, a specific focus on women during implementation is necessary.

The implication of the social dimension of agroecology is tightly linked to the achievement of the United Nations Sustainable Development Goals (SDGs). Agroecology promotes sustainable and resilient agricultural practices and therefore allowing a number of specific SDGs (FAO 2018d; Caldwell and Wang 2020):

- **No Poverty (SDG 1):** Agroecology supports smallholder farmers by reducing production costs and increasing income stability through sustainable practices.
- **Zero Hunger (SDG 2)**: By optimizing the use of local resources and enhancing biodiversity, agroecology improves food security and nutrition.
- **Good Health and Well-Being (SDG 3)**: Minimizing the use of harmful agrochemicals, agroecology promotes healthier ecosystems and reduces health risks for farmers and consumers.
- **Quality Education (SDG 4):** Agroecology emphasizes knowledge sharing and education, empowering farmers with practical and locally adapted knowledge.
- **Gender Equality (SDG 5)**: Women play a crucial role in agroecological practices, and these systems can advance women's rights and empowerment.
- **Clean Water and Sanitation (SDG 6)**: Agroecology promotes efficient water use and prevents water pollution, contributing to sustainable water management.
- **Decent Work and Economic Growth (SDG 8)**: By creating new employment opportunities and enhancing the resilience of rural livelihoods, agroecology supports sustainable economic growth.
- **Responsible Consumption and Production (SDG 12)**: Agroecological practices encourage sustainable production methods and responsible consumption patterns.
- **Climate Action (SDG 13)**: Agroecology enhances the resilience of farming systems to climate change by promoting biodiversity and sustainable land management.
- **Life on Land (SDG 15)**: By fostering biodiversity and sustainable land use, agroecology helps protect and restore terrestrial ecosystems.

Agroecology's holistic approach integrates ecological principles with social equity, making it a powerful tool for achieving multiple SDGs.

4.4 Political Dimensions of Agroecology Associated with Food Sovereignty

Agroecology places the needs and interests of small-scale food producers at the forefront. These small-scale producers supply the majority of the world's food, yet have long been overshadowed by large industrial food and agricultural systems. By prioritizing them, agroecology challenges the existing power structure in the food industry. It aims to transfer the control of key resources such as seeds, biodiversity, land, water, knowledge, and the commons into the hands of those involved in the food system. This transfer of control enables more integrated and sustainable resource management. Agroecology also has the potential to reshape power relationships. It encourages greater participation of food producers and consumers in decision-making processes related to food systems. This not only gives them a voice but also paves the way for new governance structures that are more inclusive and democratic.

Furthermore, agroecology requires a supportive policy environment. This includes favourable public policies, policymakers, and institutions that back its development and sufficient public investment. It also promotes forms of social organization that are essential for the decentralized governance and local adaptive management of food and agricultural systems. It stimulates self-organization and collective management at various levels, from local farmer organizations to global networks of different stakeholders like consumers, research organizations, and academic institutions.

As Cuéllar and Sevilla (2009) pointed out, food sovereignty represents the political aspect of agroecology. This implies that food sovereignty is an integral part of the fight against the dominance of large-scale agribusiness. The present predominant agricultural model marginalizes farmers, promotes industrial monoculture, encourages land grabbing, and leads to the contamination of soil through the use of overuse of chemicals.

The principles of agroecology often conflict with capitalist large-scale agribusiness due to their fundamentally different approaches to agriculture and food systems. Here are some key points of contention:

- **Profit vs. Sustainability:** Prioritization of profit often leading to practices that can harm the environment, such as monocropping and heavy use of chemical inputs. Agroecology, however, emphasizes sustainability, biodiversity, and ecological balance (Wezel et al. 2020).
- **Scale and Control:** Large-scale industrial farming is controlled by a few corporations, which can lead to dependency on external inputs and market fluctuations. Agroecology promotes small-scale, diversified farming systems that are locally controlled and reduce dependency on external inputs.
- **Market Dependency:** Dependency on global commodity markets can be volatile and unfair to small farmers. Agroecology seeks to build local food systems that are resilient and less dependent on global markets.

- **Social Equity:** A concentration of wealth and resources in the hands of a few exacerbates social inequities. Agroecology emphasizes social values, fairness, and participation, aiming to create more equitable food systems.
- **Knowledge and Innovation:** Agricultural knowledge and innovation have become proprietary and often controlled by corporations. Agroecology values the co-creation and sharing of knowledge among farmers, scientists, and communities.

These conflicts highlight the challenges in integrating agroecological principles within the existing agri-food framework, as they often require a shift in values and practices towards more sustainable and equitable food systems.

While agroecology and communism share some common goals, such as promoting social equity and sustainable practices, there are still areas where they might conflict:

- **Centralized Control vs. Local Autonomy:** Communism often involves centralized planning and control over agricultural production. Agroecology, however, emphasizes local autonomy and community-based management of resources (Dumont et al. 2021).
- **Uniformity vs. Diversity**: Communist agricultural policies have historically promoted large-scale, uniform farming practices to ensure food security and efficiency. Agroecology, in contrast, values biodiversity and diverse farming systems tailored to local ecological conditions (Nishikawa and Pimbert 2021).
- **Innovation and Knowledge Sharing:** In a communist system, agricultural innovation and knowledge might be controlled by the state. Agroecology promotes the co-creation and sharing of knowledge among farmers, scientists, and communities, which can be at odds with centralized control (Nishikawa and Pimbert 2021).
- **Economic Models:** While communism seeks to eliminate private ownership and create a classless society, agroecology can operate within various economic models, including those that allow for private ownership and market-based exchanges (Dumont et al. 2021).

In sum, the political dimension of agroecology represents a significant shift in power within food systems. It moves the focus away from the interests of a small number of large-scale industrial agricultural entities and towards direct producers. By doing so, it helps to rectify the injustices inflicted by corporate dominance in the current food system. When incorporated into a food sovereignty approach, agroecology becomes a vehicle for democratic transition in food systems. It empowers various groups such as peasants, pastoralists, fisherfolks, indigenous peoples, and consumers, allowing their voices to influence policy-making at community, national, and international levels. This, in turn, enables these groups to claim and achieve their right to food. Agroecology movements, often led by grassroots food producers and consumers, play a crucial role in spreading agroecological practices to other farmers and communities. This horizontal scaling-up is complemented by the need for vertical scaling-up, which depends on a conducive public policy environment. A favourable policy environment can multiply agroecological solutions, facilitating the widespread adoption of agroecology and contributing to food security, climate resilience, and sustainable development.

References

Alaniz, E. P., and N. M. Dal Ri. 2019. Educação, trabalho e formação profissional no MST: a experiência do Curso de Agroecologia da Escola Milton Santos. *Revista Brasileira de Educação do Campo* 4:e6789. https://doi.org/10.20873/uft.rbec.v4e6789.

Altieri, Miguel A. 1983. *Agroecology*. Berkeley, CA: University of California Press.

Altieri, Miguel A. 1987. *Agroecology: the scientific basis of alternative agriculture*. Boulder, CO: Westview Press.

Altieri, Miguel A. 2000. *Agroecology: principles and strategies for designing sustainable farming systems*. New York: Hayworth press. Source: College of Natural Resources: University of California, Berkeley.

Altieri, M. A., and P. Rosset. 1995. Agroecology and the conversion of large-scale conventional systems to sustainable management. *International Journal of Environmental Studies* 50: 165–185.

Anderson, F. 2018. *Food sovereignty now! European coordination via Campesina*. https://via-campesina.org/en/wp-content/uploads/sites/2/2018/02/Food-Sovereignty-A-guide-Low-Res-Vresion.pdf. Accessed May 5, 2025.

Azzi, G. 1928. *Agricultural ecology (in Italian), Edition*, 237. Turin: Tipografia Editrice Torinese.

Azzi, G. 1942. *Agricultural ecology (in Italian)*, 204. Città di Castello: Edition Dante Aleghieri.

Azzi, G. 1956. *Agricultural ecology*, 424. London: Constable & Company.

Benzin, B.M. 1925. *Agroecological characteristics description and classification of the local corn varieties chorotypes*. Book, (Publisher unknown so far, B. M. Bensin was born in Russia on 30 January 1881. He received his education in agricultural sciences at the University of Minnesota, graduating with an M.S. in 1912.)

Borsatto, Ricardo Serra, and Vanilde F. Souza-Esquerdo. 2019. MST's experience in leveraging agroecology in rural settlements: lessons, achievements, and challenges. *Agroecology and Sustainable Food Systems* 43 (7–8): 915–935. https://doi.org/10.1080/21683565.2019.1615024.

Caldwell, C. D., and S. L. Wang, eds. 2020. *Introduction to agroecology*. Singapore: Springer.

Carroll, C. R., J. H. Vandermeer, and P. Rosset, eds. 1990. *Agroecology*. New York: McGraw-Hill Publications.

CIDSE (Coopération Internationale pour le Développement et la Solidarité). 2018. *The principles of agroecology. Towards just, resilient and sustainable food systems*. Brussels.11 pp.https://www.cidse.org/publications/just-food/food-and-climate/the-principles-of-agroecology.html. Accessed May 5, 2025.

Conway, G. R. 1985. Agroecosystems analysis. *Agricultural Administration* 20:31–55.

Cox, G. W., and M. D. Atkins. 1979. *Agricultural ecology: an analysis of world food production systems*. San Francisco: W.H. Freeman.

Cuéllar, M., and E. Sevilla. 2009. Aportando a la construcción de la Soberanía Alimentaria desde la Agroecología. *Ecología Política* 38:43–52.

de Schutter, Olivier. 2011a. *Agroecology and right to food*. Report presented at the 16th Session of the United Nations Human Rights Council[R/OL]. http://www.srfood.org/images/stories/pdf/officialreports/20110308_a-hrc-16-49_agroecology_en.pdf. Accessed June 17, 2024 in Canada.

de Schutter, Olivier. 2011b. *Eco-agriculture (agroecology) can double food production in 10 years*. http://www2.ohchr.org/english/issues/food/index.htm (cited from Stephen R. Gliessman 2011. Editorial: agroecology and the growing food crisis. J Sustain Agric. 35:697–698).

de Schutter, Olivier. 2012. Agroecology, a tool for the realization of the right to food. In *Agroecology and strategies for climate change*, Sustainable agriculture reviews, ed. E. Lichtfouse, vol. 8. Dordrecht: Springer. https://doi.org/10.1007/978-94-007-1905-7_1.

de Schutter, Olivier. 2014. *Agroecology and the right to food.* Interview by Margriet Goris. http://leisaindia.org/wp-content/uploads/2014/07/LEISA-India-June-2014-Page-18-19.pdf. Accessed March 24, 2025.

Dumont, Antoinette M., Gaëtan Vanloqueren, Pierre M. Stassart, and Philippe V. Baret. 2016. Clarifying the socioeconomic dimensions of agroecology: between principles and practices. *Agroecology and Sustainable Food Systems* 40 (1): 24–47.

Dumont, Antoinette M., Ariani C. Wartenberg, and Philippe V. Baret. 2021. Bridging the gap between the agroecological ideal and its implementation into practice. A review. *Agronomy for Sustainable Development* 41:32. https://doi.org/10.1007/s13593-021-00666-3.

Food and Agriculture Organization of the United Nations (FAO). 1996. *Rome Declaration on World Food Security and World Food Summit Plan of Action.* https://www.fao.org/3/W3613E/W3613E00.htm. Accessed July 6, 2024.

Food and Agriculture Organization of the United Nations (FAO). 2018a. *2nd International Symposium on Agroecology: Scaling up agroecology to achieve the Sustainable Development Goals (SDGs).* Chair's summary. Rome, Italy: FAO.

Food and Agriculture Organization of the United Nations (FAO). 2018b. *The 10 elements of agroecology: guiding the transition to sustainable food and agricultural systems.* Rome. http://www.fao.org/3/i9037en/i9037en.pdf. Accessed April 29, 2025.

Food and Agriculture Organization of the United Nations (FAO). 2018c. *Agroecology Knowledge Hub. The 10 elements of agroecology.* Rome. http://www.fao.org/agroecology/knowledge/10-elements. Accessed April 29, 2025.

Food and Agriculture Organization of the United Nations (FAO). 2018d. *Agroecology Knowledge Hub. Agroecology and the Sustainable Development Goals (SDGs).* Agroecology and the Sustainable Development Goals | Agroecology Knowledge Hub | Food and Agriculture Organization of the United Nations. Accessed May 6, 2025.

Francis, C., G. Lieblein, S. Gliessman, T. A. Breland, N. Creamer, Salomonsson L. Harwood, J. Helenius, D. Rickerl, R. Salvador, M. Wiedenhoeft, S. Simmons, P. Allen, M. Altieri, C. Flora, and R. Poincelot. 2003. Agroecology: the ecology of food systems. *Journal of Sustainable Agriculture* 22:99–118.

Gliessman, Stephen R., ed. 1990. *Agroecology: researching the ecological basis for sustainable agriculture*. New York: Springer.

Gliessman, Stephen R. 1998. *Agroecology: ecological processes in sustainable agriculture.* Chelsea, MI: Ann Arbor Press.

Gliessman, Stephen R. 2020. Transforming food and agriculture systems with agroecology. *Agriculture and Human Values* 37:547–548.

HLPE (the High-Level Panel of Experts on Food Security and Nutrition). 2019. *Agroecological and other innovative approaches for sustainable agriculture and food systems that enhance food security and nutrition.* A report by the High-Level Panel of Experts on Food Security and Nutrition of the Committee on World Food Security, Rome. http://www.fao.org/3/ca5602en/ca5602en.pdf. Accessed May 25, 2025.

Klages, K. H. W. 1928. Crop ecology and ecological crop geography in the agronomic curriculum. *Journal of American Society of Agronomy* 10:336–353.

Klages, K. H. W. 1942. *Ecological crop geography*, 615. New York: Macmillan Company.

Lagier, Claire. 2020. Constructing legitimacy? Agroecology within and beyond the Brazilian Landless Workers' Movement (MST). *Environmental Education Research* 26 (2): 299–300. https://doi.org/10.1080/13504622.2019.1687650.

Leopold, Aldo. 1966. *A sand country almanac with essays on conservation from round river.* New York: Ballantine Books.

Lucantoni, Dario. 2020. Transition to agroecology for improved food security and better living conditions: case study from a family farm in Pinar del Río, Cuba. *Agroecology and Sustainable Food Systems* 44 (9): 1124–1161. https://doi.org/10.1080/21683565.2020.1766635.

LVC. 2007. *Declaración del Foro Mundial sobre la Soberanía Alimentaria.* Nyéléni, Mali: La Vía Campesina.

Nicholls, C., M. A. Altieri, and L. Vazquez. 2016. Agroecology: principles for the conversion and redesign of farming systems. *Journal of Ecosystem and Echography* S5:010. https://doi.org/10.4172/2157-7625.S5-010.

Niggli, Urs, Martijn Sonnevelt, and Susanne Kummer. 2021. Pathways to advance agroecology for a successful transformation to sustainable food systems. In *Science and innovations for food systems transformation*, ed. Joachim von Braun, Kaosar Afsana, Louise O. Fresco, and Mohamed Hag Ali Hassan. Cham: Springer.

Nishikawa, Yoshiaki, and Michel Pimbert. 2021. *Seeds for diversity and inclusion: agroecology and endogenous development*. Cham: Springer. https://doi.org/10.1007/978-3-030-89405-4.

Rice, Anika M., Nathan Einbinder, and Claudia Irene Calderón. 2023. With agroecology, we can defend ourselves': examining *campesino* resilience and economic solidarity during pandemic-era economic shock in Guatemala. *Agroecology and Sustainable Food Systems* 47 (2): 273–305. https://doi.org/10.1080/21683565.2022.2140378.

Rosset, Peter Michael, Braulio Machín Sosa, Adilén María Roque Jaime, and Dana Rocío Ávila Lozano. 2011. The Campesino-to-Campesino agroecology movement of ANAP in Cuba: social process methodology in the construction of sustainable peasant agriculture and food sovereignty. *Journal of Peasant Studies* 38 (1): 161–191. https://doi.org/10.1080/03066150.2010.53858.

Schiller, Katharina J. F., Laurens Klerkx, P. Marijn Poortvliet, and Wendy Godek. 2019. Exploring barriers to the agroecological transition in Nicaragua: a technological innovation systems approach. *Agroecology and Sustainable Food Systems*. https://doi.org/10.1080/21683565.2019.1602097.

Sinclair, Fergus, Alexander Wezel, Cheikh Mbow, Susan Chomba, Valentina Robiglio, and Rhett Harrison. 2019. *The contribution of agroecological approaches to realizing climate-resilient agriculture*. Rotterdam and Washington, DC. 4. https://gca.org/wp-content/uploads/2020/12/TheContributionsOfAgroecologicalApproaches.pdf. Accessed April 25, 2025.

Stephen R, Gliessman. 2007. *Agroecology: The ecology of sustainable food systems (2nd ed.).* Boca Ratón: CRC Press.

Vamdemeer, John. 1995. The ecological basic of alternative agriculture. *Annual Review of Ecology and Systematics* 2:201–224.

Varghese, Shiney. 2022. *What is agroecology, and how can it provide solutions to the crises that plague our food systems and create food democracies?* https://www.iatp.org/agroecology-takes-center-stage-global-agenda-transforming-agriculture-and-food-systems. Accessed April 27, 2025.

Wang, S. L., C. D. Caldwell, and W. F. Zhu. 2010. Low carbon agriculture: origins, principles and strategies (in Chinese). *Research of Agricultural Modernization* 31 (5): 604–607.

Wezel, Alexander, Stéphane Bellon, T. Dore, C. Francis, D. Vallod, and C. David. 2009. Agroecology as a science, a movement and a practice. A review. *Agronomy for Sustainable Development* 29:503–515.

Wezel, Alexander, Margriet Goris, Janneke Bruil, Georges F. Félix, Alain Peeters, Paolo Bàrberi, Stéphane Bellon, and Paola Migliorini. 2018. Challenges and action points to amplify agroecology in Europe. *Sustainability* 10:1598. https://doi.org/10.3390/su10051598.

Wezel, Alexander, Barbara Gemmill Herren, Rachel Bezner Kerr, Edmundo Barrios, André Luiz Rodrigues Gonçalves, and Fergus Sinclair. 2020. Agroecological principles and elements and their implications for transitioning to sustainable food systems. A review. *Agronomy for Sustainable Development* 40:40. https://doi.org/10.1007/s13593-020-00646-z.

Chapter 10
Eco-Agriculture: Adapting to Global Climate Change by Applying Agroecology

Abstract The first nine chapters of this book established the necessity for a worldwide agroecological transition away from conventional agriculture. This final chapter applies the principles of agroecology to create a new model of inclusive agriculture, namely ecological agriculture or eco-agriculture, or in essence, the model of sustainable agriculture. The evolution of perceptions of ecological agriculture both in the West and the East are addressed. In response to climate change and other destructive environmental consequences of conventional agriculture, this chapter describes the four principles of agroecological engineering: 'integrity, coordination, circulation and regeneration'. This is expanded with the three macro-pictures of eco-agriculture, its three main missions, three technology categories, and three policy implications.

Keywords Agroecology · Eco-agriculture · Global climate change · Agroecological compensation · Eco-labelled product · Community Supported Agriculture

Learning Objectives
After reading this chapter, student should be able to:

1. Explain the concept and need for eco-agriculture.
2. Discuss how eco-agriculture can be a positive response to climate change?
3. Discuss how the following can be used as economic leverages that support eco-agriculture landscapes:
   (a) Agroecological compensation
   (b) Eco-labelled product
   (c) Community supported agriculture

****

S. Wang, C. D. Caldwell, *Advanced Agroecology*,
https://doi.org/10.1007/978-3-032-08689-1_10

> I don't trust any of the food I eat, but I just eat it and hope for the best. I worry for my children, but it's what there is here in this country [China].
>
> —Mr. Feng, interviewed in July 6, 2014 (cited from Regnier-Davies 2015, p. 38)

> Saving our planet, lifting people out of poverty, advancing economic growth… these are one and the same fight. We must connect the dots between climate change, water scarcity, energy shortages, global health, food security, and women's empowerment. Solutions to one problem must be solutions for all.
>
> —Ban Ki-moon

The greatest challenge of our era is climate change. Intensive, extractive human activities for economic development have produced imbalances in the matter and energy cycles of the biosphere. The generation and release of carbon dioxide, water vapour, methane, nitrous oxide, and other greenhouse gases (GHGs) are heating the Earth's surface. In the past hundred years, the average temperature of the Earth's surface has been increased by 1.5 °C. The warming Earth has a devastating consequence in terms of biodiversity maintenance and, ultimately, threatens human existence.

Industrial agriculture is a major source of GHGs, for instance, agriculture contributes 25% of atmospheric carbon dioxide, including releasing carbon dioxide by farming and land use pattern changes and using fossil energy to produce fertilizer, pesticide, and other inputs. Agriculture contributes 70% of the nitrous oxide, which is an offshoot of nitrogen fertilizer application, and contributes 40% of the methane both from rice paddy fields and animal husbandry. Thus, there is another name for modern conventional agriculture: high-carbon agriculture. In response, a low-carbon agriculture (LCA) or climate-smart agriculture (CSA) has been proposed to reverse the degenerating situation (Wang et al. 2010; Scholes 2012). The basic concept of LCA or CSA is to use the principles of agroecology to re-balance the carbon cycle in food systems. This approach is called ecological agriculture (hereinafter referred to as eco-agriculture). This chapter addresses the principles and technological systems of eco-agriculture as a response to climate change and, as such, is an applicable and integral part of agroecology.

1 The Concept and Need for Eco-Agriculture

Historically, eco-agriculture was first coined in 1970 by an American soil scientist, William A. Albrecht as one of the alternative agricultural approaches to modern fossil-based agriculture. This was in parallel with another three kinds of possible alternatives, i.e. organic agriculture, biological dynamic agriculture, and natural farming. The original intention was to apply principles of ecology to maintain soil structure and fertility (Mercier 1978), and it was put into practice in some states and provinces in North America. In 1981, a British agronomist Martha Kiley-Worthington formally defined eco-agriculture as (Kiley-Worthington 1981):

> The establishment and maintenance of an ecologically self-sustaining low input, economically viable, small farming system managed to maximize production without causing large or long-term changes to the environment or being ethically or aesthetically unacceptable.

Continuing to explore the principles of ecology, Worthington's definition has expanded from a focus mainly on soil to a broader focus on whole agroecosystems, targeting an appraisal of large-scale farming in Europe and America pursuing the goal of holistic balance in agroecosystem. Such approaches to eco-agriculture were gradually implemented in countries like Mexico and Brazil, some parts of Canada, especially Prince Edward Island and Saskatchewan; some European countries like England, Spain, France, Bulgaria, Republic of Lithuania, Greece, Turkey, Romania, Hungary, Poland, and Slovakia (Wang and Shi 2023; Némethová 2010; Bernal 2011; Brodt et al. 2006) and African countries like South Africa (Chitakira et al. 2022). However, in many developed countries, the term eco-agriculture was seldom mentioned after 1990 until the early of 2000s, when sustainable agriculture was the term proposed following the definition of sustainable development formally defined by the UN Brundtland Commission 1987 to cover all the abovementioned four kinds of alternative agricultures.

However, in China, there was a totally different starting point from the West. China's recent history is one of the small family farm-based production systems with lower external input and outputs. However, since the 1970s, because of huge population pressures, the national agricultural development strategy included the implementation of 'the Four Modernizations, mechanization, irrigation, electrification and chemicalization in Agriculture'. Since then, a panel of scientists mostly ecologists and agronomists, worried about the environmental consequence of this approach, convened a conference to introduce the thinking and subject of ecology and agroecology. Subsequently, Chinese eco-agriculture was proposed in the conference and was partially practised in the 1980s. Professor Ye Qianji, who first brought eco-agriculture in China, proposed the general meaning of eco-agriculture in 1982 (Ye 1982, 1988):

> In short, the future of agriculture requires that those who dominate everything in the agricultural ecosystem must follow the laws of nature and economics, based on the recent past, look to the future, promote and maintain a virtuous cycle, and create an ideal ecosystem that often maintains the best balance for our generation and future generations, which we can call efficient ecosystem, that is, ecological agriculture.

Then, Ye specifically defined the concept of Chinese eco-agriculture (Ye 1988; Ye and Luo 1987):

> Starting from the systematic approach, following the principles of ecology, economics, and eco-economics, adhering to the natural ecological laws and social economic laws, utilizing modern scientific and technological means and management techniques, drawing upon the essence of traditional agriculture and modern agriculture, through the composite cyclic mechanism of ecological systems and economic systems, a modern agricultural production model is established that features multiple objectives, functions, components, and multiple levels of a well-combination, structured, open, cyclic, and exchangeable system with harmonious relationships and coordinated development.

Under those guiding instructions, a series of top-down eco-agriculture building projects were implemented, and a series of eco-agriculture models were built during the period between 1982 and 2002 (Song et al. 2022). These included the Liuming Ying eco-village Dou Dian eco-village in Beijing Municipality, Teng Tou eco-village in Zhejiang province, Xiao Zhang Zhuang eco-village in Anhui Province, Qian Wei eco-village in Shanghai Municipality, Tie Xi eco-village in Heilongjiang Province, Dafeng eco-county in Jiangsu Province, Chongming eco-county in Shanghai Municipality, Dazu eco-county in Sichuan Province and Simao eco-county in Yunnan Province (Sanders 2000), Jingshan county of Hubei province (Shi and Gill 2005), and Yilan and Hualien counties in Taiwan provinces (Chen et al. 2021). Despite these successes, there were arguments in academic arenas on what will represent China's green agriculture, eco-agriculture, and sustainable agriculture. It was concluded that eco-agriculture would be an applicable model of sustainable agriculture in China, while the latter was more an ideal metaphor than applicable practice (Wang and Lin 1999). With this background, eco-agriculture development in China progressed quickly. According to China's Ministry of Agriculture, there were 370 eco-agriculture models screened nationwide, and 50 eco-counties were being built during 1983–2003 (Shi 2003), and another 51 eco-counties were constructed during 2003–2008 (Song et al. 2022). Chinese eco-agriculture now has a concluding definition as (Wang and Chen 2009):

> An agricultural system that combines modern scientific achievements with the essence of traditional agricultural technology by applying the principles of agroecology and systematic scientific methods to achieve economic viability and ecological rationality.

However, due to the loss connection with original definition by Martha Kiley-Worthington in 1981 which was considered not matching to China's macro-economic scenario of agriculture development at the time (Wang and Shi 2023), the booming eco-agriculture in China has seldom been introduced in Western academia as an ecological application, if have, it was named as agroecological engineering (Lescourret et al. 2015; Sun and Qi 2017).

Meanwhile, eco-agriculture was re-known internationally with two cornerstone recognitions.

Just prior to the new millennium, the first stage of an eco-agriculture renaissance emerged. This was the result of a deepening understanding of the environmental impacts of agriculture and the popularization of the concept of sustainable development. Eco-agriculture emerged as a new agriculture model aimed at the dual purposes of providing ecologically compatible food production and maintaining biodiversity (McNeely and Scherr 2001). Mostly in this case, eco-agriculture was the synonym for organic agriculture in east European countries like Romania, Republic of Moldova, and Slovakia (Sandu et al. 2023; Ursu and Petre 2022; Patarlageanu et al. 2022; Jitareanu et al. 2022; Némethová 2010; Teodorescu 2008; Larisa and Tatiana 2008; Sriskandarajah et al. 2006), because the latter is evidenced to be beneficial to wildlife residency while producing human food, even to be equated with biological agriculture (Cornelia et al. 2008). As such, the idea of eco-agriculture was grown out of the contradict between the lands for agricultural expansion and biodiversity conservation locally and globally and hence was expected to be a pattern of sustainable development of agriculture (Wang and Lin 1999; Patarlageanu et al. 2022; Ma and Zhuge 2024). Scientists such as Jeffrey McNeely, a conservationist, and Sara Scherr, an agricultural economist, played crucial roles in its renaissance. After the publication of McNeely and Scherr's book entitled *Eco-agriculture: Strategies to Feed the World and Save Wild Biodiversity* in 2002 (McNeely and Scherr 2002), the concept of eco-agriculture gained more attention. Subsequently, networks like eco-agriculture Partners (https://ecoagriculture.org) were formed, and international conferences and expositions on eco-agriculture were held, which promoted the exchange of knowledge and the expansion of practices related to eco-agriculture. In a sense, the concept of eco-agriculture emerged as a response to the challenges posed by the complex relationship between agriculture, rural poverty, and biodiversity. Historically, agriculture and wild biodiversity had a close yet evolving relationship. As agriculture expanded, natural habitats were converted for agricultural use, leading to significant biodiversity loss. This conversion not only affected the distribution and status of wild species but also disrupted ecological processes like pollination and seed dispersal. The term 'eco-agriculture' represents an innovative approach that aims to integrate biodiversity conservation into agricultural development. It evolved from the ecosystem approach, taking into account the needs of wild species in agricultural systems.

With recent scientific and institutional advances, eco-agriculture has developed into a set of strategies. These strategies include creating biodiversity reserves that benefit local farming communities, developing habitat networks in non-farmed areas, increasing farm productivity to reduce land conversion, minimizing agricultural pollution, modifying the management of soil, water, and vegetation resources, and mimicking natural ecosystems in farming systems. Overall, eco-agriculture offers a potential solution to simultaneously address the goals of feeding the world and saving wild biodiversity (Scherr and McNeely 2007).

As praxis domain of scaling-up agroecology being ecology of sustainable agriculture (Gliessman 1998), eco-agriculture has been broadening to be an approach to land use that aims to balance competing objectives such as enhancing agricultural productivity, ensuring ecological sustainability, and promoting rural vitality. As such, it is not a single practice or a specific farming system but rather a collection of methods that integrate biodiversity conservation into agricultural development. In the new paper, Scherr and McNeely proposed that eco-agriculture should be regarded as a comprehensive set of resource management strategies (Scherr and McNeely 2008). These strategies are designed to produce more food while preserving ecosystem services, with a special emphasis on wild biodiversity. Specific strategies include creating biodiversity reserves that benefit local farming communities, developing habitat networks in non-farmed areas, increasing farm productivity to reduce land conversion for agriculture, minimizing agricultural pollution, modifying the management of soil, water, and vegetation resources, and mimicking natural ecosystems in farm systems (Buck et al. 2004, 2006). These strategies cover various aspects of agricultural production and land management, aiming to achieve a harmonious co-existence between agriculture and the environment. Overall, the stage of eco-agriculture could be summarized as follows (Cheatham III 2022; Buck et al. 2006):

- **Conceptual Development**: Ecological agriculture stands in contrast to industrial agriculture. It emphasizes integrating ecological processes into farming, following principles such as biospecies, equality, limited growth, and non-dominating science. It promotes a view that nature knows best, advocating that humans relinquish excessive control and seek answers from nature. For example, it opposes the use of tilling, herbicides, pesticides, GMOs, and antibiotics, aiming to achieve a harmonious co-existence between humans and the Earth through mutual reciprocity (Altieri 2004). Eco-agriculture focuses on improving human management of land and natural resources, striving to achieve three core goals simultaneously. Firstly, it conserves and maintains native biodiversity and ecosystem services. Secondly, it ensures sustainable agricultural production to guarantee a stable supply of agricultural products and services. Thirdly, it provides viable livelihoods for local people, enhancing their quality of life. These three goals are interrelated and jointly define the development direction of eco-agriculture.
- **Practical Development**: Farmers who engage in eco-agriculture in this stage usually practice methods like diverse planting and raising. They focused on the symbiotic relationships between different species, such as the 'three sisters' planting method (corn, beans, and squash) that benefits the soil and plant growth. They also pay attention to soil health, for instance, adding cover crops to improve soil quality. Additionally, some eco-agriculture farms have transitioned to subscription-based CSA models, strengthening the connection between producers and consumers and enhancing community support for local agriculture. With expansion of agroecology as the ecological basis of sustainable agriculture, the practices of eco-agriculture involved multiple strategies. For example, creating biodiversity reserves benefits local farming communities. Developing habitat networks in non-farmed areas of agricultural landscapes enhances ecological connectivity. Increasing farm productivity reduces the need for new land conversion, alleviating pressure on the natural environment. Minimizing agricultural pollution protects the ecological environment. Modifying the management of soil, water, and vegetation increases natural capital. Designing farm systems to mimic natural ecosystems improves the stability and sustainability of ecosystems.

The second stage of the eco-agriculture renaissance, occurred in the early 2000's. This was the refinement of eco-agriculture as a path to low-carbon, climate- smart agriculture (Francis et al. 2003; Gliessman 2007; Wang et al. 2010). In 2011, Olivier de Schutter, the UN Food Rights committee officer, submitted a report named 'Agroecology and Right to Food' to the UN commission, which was subsequently approved by the UN (de Schutter 2010). This report used examples from the successful agroecology efforts in the global South (including Cuba, Brazil, and Mexico) and expanded the definition of agroecology as a science, a practice, and social movement Wezel et al. (2009). It was understood that agroecology will play a pivotal role in agri-food system transition from high carbon to low carbon, from industrial to ecological, from unsustainable to sustainable (Zhang 2024). Both in de Schutter's report and in the Wezel et al.'s article, agroecology, which was interpreted as a practice, refers to a kind of sustainable agriculture practice, essentially synonymous with eco-agriculture that has been practised as a model of sustainable agriculture for more than 40 years (since 1981) (Song et al. 2022). Therefore, eco-agriculture as a scientific term should be substituted for the term agroecology when it is used solely as a practice in the world's academia. Eco-agriculture can therefore be interpreted as (Wang and Shi 2023):

> A transformative agricultural system that applies the principles and methods of agroecology to rebuild an environmentally friendly, economically viable and climatically adaptive agri-food systems globally, regionally and locally.

Such an internationally applicable concept of eco-agriculture will pave the way for world agri-food systems to become truly sustainable agriculture and food systems. However, to facilitate such new models the three levels of agroecology should be explicitly defined:

Firstly, agroecology as a science needs to be developed and strengthened globally; agroecology education and research should be updated and upscaled with academic programs and a cluster of curricula in the agriculture-associated universities and colleges (Wang et al. 2019; Wang and Shi 2023).

Secondly, agroecology as a practice needs to have an explicit methodology of agroecological engineering (Lescourret et al. 2015), i.e. integrity, coordination, circulation, and regeneration, which were first addressed by Professor Ma Shijun in 1979 (Ma and Li 1987, cited from Sun and Qi 2017 and Xu et al. 2024). Each is explained as follows:

- **Integrity:** Basically, agroecology offers guidelines to production agriculture in a holistic and sustainable manner, meaning that each sector of agriculture, farming, husbandry, fishery, processing, and marketing, is connected through energy flow, matter cycle, information transfer, and value(money) transition and form an agroecosystem. Therefore, for each of the abovementioned sectors, factors should be integrated and coordinated in the design phase of eco-agriculture; otherwise, the integrity of agroecosystems will be lost.
- **Coordination:** To maintain the integrity of a certain agroecosystem, each sector of eco-agriculture needs not only to link qualitatively but also to connect quantitatively. Such quantitative connection is designed based on the principles of the food chain and ecological pyramid of ecology, letting the residue of the former sector become the resource of the latter sector of agriculture.
- **Circulation**: From the perspective of the ecological matter cycle, within an eco system, matter changes in form while maintaining a constant quantity as it circulates. Geographically, matter first flows from the environment to plants, then proceeds to various levels of consumers, and finally returns to the environment. Thus, the matter cycle has two directions: one leads to the macro-environment, that is, the five spheres, and the other stays in soil and then returns to plants to

restart the cycle. The latter is considered to be the advantage of eco-agriculture, through which eco-agriculture demonstrates its superiority.

- **Regeneration**: From the viewpoint of energy flows in ecosystems, energy is lost as trophic levels step forward in the food chain. Therefore, the trophic links of food chains in an eco-agriculture system can be cut and regenerated; e.g., the energy in plant and animal residues can be recovered via biogas and compost in an eco-agriculture system. Biogas is used for household energy, the residue of biogas used for clean fertilizer for farming, and compost on the soil. Other opportunities for closing the nutrient cycles include mixing brown and green manure to produce an organic matter enriched fertilizer substituting for synthetic fertilizer.

Thirdly, when defined as a movement, eco-agriculture (agroecology in de Schutter's report) means a social movement for rebuilding a sustainable food system with each stakeholder's involvement and participation. Most importantly, it needs bi-directional movement at the point where the top-down policy and the bottom-up movement meet (Scott et al. 2014). In China, consumers have become more aware of food safety issues since 2008 and there has been the establishment of ecological civilization as the national development strategy in 2012. This has led to the mainstreaming Chinese eco-agriculture, which has the potential to inspire the world's eco-agriculture development (Zhang 2024; Wang and Shi 2023; Buck and Milder 2011).

To sum up, as the drawbacks of conventional agriculture in various countries around the world are continuously reported and revealed, the ecological transformation of agriculture has become an irresistible trend in the post-petrochemical agriculture era. Eco-agriculture serves as the primary vehicle for the sustainable development of global agricultural and food system particularly in response to climate change (Wang and Shi 2023).

2 What Do We Mean by Eco-Agriculture in Response to Climate Change?

Humanity now faces two major challenges. First, there is too much carbon dioxide in the atmosphere. Since the beginning of the industrial era, atmospheric CO_2 concentrations have surged from 280 parts per million to nearly 430 parts per million by 2025, an increase of more than 50%. The burning of fossil fuels is the main cause, but agricultural activities are also a significant source of emissions. Each year, 13.7 billion tons of CO_2 or equivalent greenhouse gases are released into the atmosphere through agricultural processes, and more than a quarter of global greenhouse gas emissions originate from food production.

Another carbon-related problem is the lack of carbon in the soil. Intensive tillage practices are widely used to meet the needs of a growing population, but this has led to a dramatic decline in soil carbon levels worldwide. Since the dawn of agriculture, about 12,000 years ago, the amount of carbon lost from the top 2 m of soil is estimated to be about 133 billion tons (about 8% of total soil organic carbon). Of that, about a third has been lost since the Industrial Revolution in the nineteenth century. This imbalance reveals the enormous potential of agriculture: while it is currently a source of carbon emissions, it also has the potential to transform into a carbon sink, altering the planet's climate change trajectory. By supporting and enhancing the natural process of converting CO_2 into soil carbon, we can replenish soil organic carbon stocks, and this process is relatively easy to achieve (Nogrady 2024).

Climate change and agriculture are inextricably linked; industrialized agriculture is one of the most significant factors affecting climate change, and climate change in return has a huge impact on agriculture. Hence, climate change is a crucial challenge to global food security. Since agroecology deals with the ecology of the entire food systems (Francis et al. 2003), it first needs to set a goal to secure global food security under our present climate change era (Vermeulen et al. 2012). With such a priority goal, eco-agriculture needs to include the frequently mentioned low-carbon agriculture or climate-smart agriculture (CSA) in academia. It must be encouraged to be applied in two ways, adaptation and mitigation for dealing with climate change (Tomich et al. 2011). To achieve this, the following triple-four strategies of eco-agriculture should be applied to secure global food security under climate change:

2.1 *The Three Pillars of Eco-Agriculture*

As shown in Fig. 10.1, eco-agriculture that coordinates the linked challenges of global climate change and food security needs to embrace agricultural production, biodiversity conservation, and viable local livelihood, three pillars of eco-agriculture intertwined. Agricultural production should not only be maintained but also be sustainably increased by at least 50% between now and the year 2050 to respond to population growth and food demand. The demand for meat consumption increases with wealth in developing nations; this is somewhat offset by increasing trends in plant-based diets but is not yet effective in decreasing the overall demand for food. The question remains how this can happen in the face of climate change, resource depletion, and environmental degradation. Therefore, increasing biodiversity in the agroecosystem is one of the keys to eco-agriculture structure (McNeely and Scherr 2001, 2002); another key is to ensure rural livelihood; otherwise, there will be no farmers producing our food.

2.2 *The Three Main Missions of Eco-Agriculture*

The three main missions of eco-agriculture include:

- **Transforming the agricultural industry from focusing on pure economic return to supporting public life.** Due to the unbridled pursuit of crop yield,

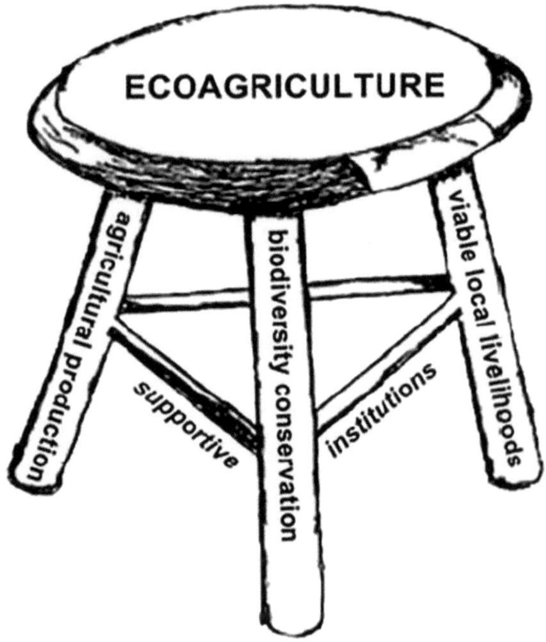

Fig. 10.1 The three pillars of eco-agriculture in terms of its climatically adaptive manner. (Source: Scherr et al. 2014, used with permission)

agricultural production and economic return, the current conventional agriculture has largely ignored livelihood issues, (rural) ecology, and agricultural multiple functionalities characterized by 'the three life' utilities: food production that supports the lives of all citizens, economic return that supports producers' daily life, and rural and agricultural habitats that accommodate and entertain people both from rural and urban area, including even wildlife (i.e. regulating and supporting services within the ecosystem service category). In this regard, agriculture can be defined as an industry that converts sunlight into people's healthy and happy life (Wang 2005; Wang et al. 2019). Such a precious life industry cannot be realized only by a single market economic mechanism but needs to be transformed into a public utility that provides public goods, allows society to participate in investment and pricing, and collaborates to achieve 'industrial prosperity' and 'rich life' in the countryside.

- **Converting agricultural production system from an ecologically vicious circle to an ecologically virtuous circle.** For a long time, the high yield of agricultural organisms has been the priority, and excessive fossil fuel-based inputs have led to the destruction of cultivated land and soil, water pollution, greenhouse gas emissions, and eventually produces a serious food safety problem, threatening consumers' health. To reverse such an ecologically vicious consequence, ecological agriculture, which is based on the concepts, principles, and methods to coordinate the agricultural, forestry, and animal husbandry into a food chain, will be the only way to bring back an 'ecological liveable' and 'rich life' to people.

- **Promoting agricultural administration from a broken chain to 'cropping husbandry-processing' and 'agriculture-industry-service' integrations**. For some time, farming, breeding, processing, trade, and services based on agricultural products have been separated geographically and in terms of business entities both due to social division of labour and to discrete disciplines and subjects in the universities and colleges. Conversely, the profit of the whole industry is very low for the farmers engaged in production, which is the root cause of the formation of rural economic problems. Consumers and producers are intentionally separated geographically and psychologically, which is the crux of the food safety problem and the lack of trust by consumers in food safety. Ecological agriculture emphasizes the integrated management of 'farming, agroforestry and husbandry' linkage and 'agriculture, industry and commerce' connection, which is a viable and feasible way to solve the problems of agricultural industry fragmentation and rural economy.

2.3 *The Three Technology Categories of Eco-Agriculture*

The three technology categories include:

- **A multidimensional land use system for fully and efficiently utilizing the spatial-temporal resources.** Multidimensional land use technology is one of the key technology systems of eco-agriculture that utilizes ecological niche and biological succession principles of ecology to develop and utilize natural resources fully and efficiently. In the vertical space, it exploits the different characteristics of crops corresponding to their different requirements for environmental factors. There is the opportunity for the organic combination of planting on land (Fig. 10.2) and aquaculture, which can set up a multiple species common habitat. This strategy of efficient use of materials and energy produces a multi-level utilization of resources. This is also known as three- dimensional farming or three-dimensional aquaculture. For example, in China, following the principles of traditional agroforestry, the emerging 'under-forest economy' uses forest resources and environment to plant crops, Chinese herbal plants, edible fungi, and even raise livestock and poultry to form a composite system of agriculture (Fig. 10.3), forestry, and animal husbandry with biodiversity, in order to fully and efficiently use natural resources, reduce the input of physical energy outside the system, and form an efficient agricultural production system.
- **A material and energy recycling system for building self-cleaning production system**. By utilizing our understanding of the ecological principles regarding the flow of matter and energy through the food chain pyramid, it is possible to design multistage systems that optimize food chain trans formation efficiency and minimize waste, providing for a sustainable production system. For example, fish farming in rice fields is one of the simplest ecological agriculture models, using the water space between rice plants to add 'fish' food links to achieve

Fig. 10.2 The ample resource space under mature forestry used for growing herb plant medicine in Hainan Island. (Photo credit: Dr. Songliang Wang, Dec. 23, 2013)

multiple goals: Firstly, in order to ensure the survival of fish, rice fields should not apply chemical fertilizers and pesticides; Secondly, fish will eat harmful organisms on rice plants, and fish manure can fertilize the field, which can reduce the input of fertilizers and pesticides. Thirdly, the movement of the fish between the rice plants promotes the nutrient reduction reaction of paddy soil, greatly reducing methane emissions.

- **An integrated pest management system for securing land and food health.** The comprehensive pest control technology system is a combination of application of cultivation technology, physical means, biological control, and rational use of pesticides by using ecological niche and biological species relationship principles of ecology to reduce crop losses caused by pests and achieve substantial pesticide reduction. Consequently, such technology applications not only maintain the health of agricultural production environment but also ensures the safe supply of food and important agricultural and sideline products for both urban and rural residents.

2.4 The Three Policy Mechanisms of Eco-Agriculture

The three policy mechanisms of eco-agriculture include:

- **Updating the theoretical system of eco-agriculture by promoting the agroecology discipline.** As has been addressed earlier in this book, ecology is the theoretical basis of eco-agriculture, and agroecology, which basically integrates

Fig. 10.3 A case of the 'under-forest economy' from Yangchuan County of China. (Photo credit: Dr. Songliang Wang, Mar. 15, 2015)

agronomy and ecology, is the basis for constructing eco-agriculture. Compared with developed countries and some foreign developing countries (such as South American countries), the construction of agricultural ecology theory in some developing countries like China is far behind the practice of eco-agriculture. To develop eco-agriculture in those countries, it is necessary to strengthen the construction of agricultural ecology discipline, scientific research investment, and professional education, to set up undergraduate majors of agricultural ecology in agricultural and forestry colleges and universities across the country and to develop a complete educational system of agricultural ecology from undergraduate to doctoral education, so as to reserve theoretical basis and corresponding talents for eco-agriculture development.

- **Fully realizing the eco-product value of ecological agriculture through the sound agroecological compensation institution.** To promote the agroecological transition of world agriculture, eco-agriculture system construction must progress. It is necessary to build a green action list for developing eco-agriculture, which mainly guides the first-line producers to understand what eco-agriculture is and how to carry out eco-agriculture. In addition, the certification system of eco-agricultural products allows the producers engaged in eco-agriculture to share the due benefits. Therefore, it is urgent to study and to build the agroecological compensation mechanism, especially the farmland eco-compensation mechanism, which includes: What kind of behaviour can be compensated for? Who gets compensation? How much is the compensation? Who will compensate? How to compensate?

- **Ensuring the effective governance of territorial resources via unifying spatial planning.** Given that the regulatory role of law is the highest guarantee for governing the healthy development of eco-agriculture, it is necessary to improve the territorial space planning as soon as possible and sort out all the existing plans, rules, regulations, and laws related to the development of eco-agriculture in each nation. It is also imperative that the current policy constraints of agricultural resources and environment on promoting the green development of agriculture improve relevant laws and regulations, form a more effective supervision mechanism for the development of eco-agriculture, and thus consolidate the legal basis for the construction of eco-agriculture worldwide.

3 Promoting Economic Leverages That Support Eco-Agriculture Landscapes

As addressed in the first section of this chapter, eco-agriculture is a land use system emerging as a crucial strategy for conserving wild biodiversity, especially in highly populated, poor rural areas. Rooted in the concept of 'ecosystem management', it aims to simultaneously produce food and safeguard wildlife and critical ecosystem services. This approach takes enhancing rural livelihoods through more productive and profitable farming systems as a core strategy for both agricultural development and biodiversity conservation.

Historically, conventional wisdom held that modern farming was largely incompatible with wildlife conservation and other key ecosystem services. As a result, environmental policies typically relied on land use segregation, establishing protected areas excluding agriculture. However, this strategy became unviable in many parts of the world, especially those with high rural population density, poverty, and dependence on farming. With the increasing awareness of the extensive human impact on land suitable for agriculture and the recognition that biodiversity conservation in these areas is crucial in the long term, eco-agriculture emerged as an alternative. It represents a shift from the traditional focus on 'trade-offs' between poverty reduction and the environment to finding synergies through scientific management. Over time, with the development of relevant scientific disciplines such as conservation biology, agricultural ecology, and the advancement of technologies like new monitoring methods and crop breeding techniques, eco-agriculture has gradually evolved and been applied in various regions around the world, demonstrating positive impacts on wildlife populations, farm yields, and farmer incomes.

One of the most significant challenges for eco-agriculture is the absence of sufficient incentives for its development, particularly when it comes to encouraging farmers to adopt appropriate eco-agricultural practices. In regions like North America and Europe, the relatively high levels of farm subsidies (though these are gradually being reallocated towards conservation efforts) are a historical legacy. This legacy is a reflection of the small proportion of the population engaged in farming, as well as the economic and political capabilities of high-income countries to afford such subsidies.

This model of public subsidy is hard to justify in high-income countries. Moreover, it is highly unsuitable for low-income countries, where public funds have much more urgent and critical needs. Instead, public funding for eco-agriculture should be focused on areas such as research, ecosystem planning and monitoring, providing technical assistance services, and offering support to local farming and conservation organizations. This support enables these organizations to adapt eco-agriculture to local circumstances. In fact, more emphasis should be placed on phasing out subsidies for agricultural systems and inputs that incentivize practices detrimental to biodiversity.

As addressed in Chap. 4, worldwide, the rural economy has become increasingly difficult both for farmers and for many countries, due to the decreasing proportion of income sharing that accrues to farmers. Ecologists in academia are providing new arguments to economists and politicians to reverse the situation; they have developed the concept of ecosystem services to demonstrate the multiple functionalities of agriculture (Erlich and Erlich 1981, and see Chap. 3). This provides evidence for valuing in the marketplace attributes other than simply products; this includes the other parts of agriculture's multiple functions, i.e. provisioning services, supporting services, cultural services, and regulating services. One example of regulating services is soil conservation value which occupies nearly 70% of the services of agroecosystems according to pioneering research in Fujian province of China in 2011 (Zhu 2011). These arguments have mostly fallen on deaf ears and have not resulted in significant changes worldwide to agricultural value chains. Therefore, there is a need to work towards the monetization of the agroecosystem services, to properly compensate farmers for practising eco-agriculture. Such an approach has been broadly called Payment for Ecosystem Service (FAO 2007; Scherr and McNeely 2008; Chen et al. 2021), or in China, 'Valuing the ecological product'. Practically, there are three kinds of leverages that currently could be explored: agroecological compensation, eco-labelled product, and community supported Agriculture.

3.1 Agroecological Compensation

Farmers who move towards eco-agriculture are met with increasing costs, especially during the transition period from conventional to eco-agriculture. Given the current unfavourable economics on certificating agriculturally eco-products such as grains, milk, and fruits that are more dependent on natural conditions (Némethová 2010), there is a need to provide encouragement and stimulation for such farmers.

To encourage the eco-agriculture, particularly at its initiation stage, it needs to have institutional arrangements for financial compensation. Providing such farmers who practice eco-product production is termed a kind of 'payment for ecological services' (PES) (Wang and Shi 2023; Fallas Henriquez and Molina-Murillo 2017). The process for doing this is complicated and needs collaboration among scientists, policymakers, and farmers. Chapter 3 has given a detailed explanation and design on agroecosystem service assessment and payment design directly to farmers; however, agroecological compensation is more practical than scientific research and planning design. It is based on ecology but needs close links to practical, local eco-economic landscapes. This is only possible after investigating multiple options for payment for ecosystem services that are applicable to the specific situation. Consider the agroecological compensation standard of China's rice-fish eco-agriculture as a case (Moucheng et al. 2014), by using the method of farmer's willingness to accept, the latter wished to get 7462 Yuan/ha/a from both government and market; however, the provisioning services of the same year was 7393 yuan/ha/a, including the recognition of the market at the time, thus government should have paid the margin value of 69 Yuan/ha/a to meet the farmer's willingness, the number is the agroecological compensation standard at the time. It should be noted that such standards vary with the time when eco-agriculture develops and is recognized.

3.2 *Eco-Labelled Product*

Although the market usually reacts slowly, it still has instruments to facilitate the certification of eco-products. Eco-labelling and creating eco-product brands are two tested and effective options. For example, in Nicaragua, a brand for agroecological products has been developed by the Group for the Promotion of Ecological Agriculture; five products within the brand category are sold in higher price markets and stores in the larger cities and municipalities (Schiller et al. 2019). In Lishui city of Zhejiang province of China, a public brand of agroecological product named 'Lishui Shangeng' (Liao 2020) is also available to consumers. Liao Feng argues that the value of agroecological products includes the value of supporting services (such as raw material for producing agroecological products), regulating services, cultural services of Lihui Agroecosystems, which will be characterized by the Mountainous Farming of Lihui. In the same way, Pingnan County of Fujian province of China issued a public agroecological product brand 'Pingnan 800' (Fig. 10.4)， in which '800' was used to indicate Pingnan territory's elevation, describing a comparative weather condition under which the products are produced. The intent is for these products to have an advantage in marketing strategy with higher price return. In the past two decades, Pingnan has taken such advantage to produce cauliflowers as a 'Stagger the launch date' vegetable brand earning renown in city markets, but with the use of intensive chemical input to the soil and food pesticide residue. The creation of 'Pingnan 800' aims at eco-agriculture as a mandate pathway to produce agroecological product and thus to get more price recognition and return for the

Fig. 10.4 ‘Pingnan 800’ eco-product brand formally issued at November 24, 2023. (Photo credit: Dr. Songliang Wang, Nov. 24, 2023)

farmers. At the same time, the question of quality guarantee remains, business is business, following the marketing, agroecological products need to have certification to differentiate it from industrial agriculture.

3.3 Community Supported Agriculture

To decrease discrimination against farmers and to make value chains more equitable and cooperative, at the same time providing assurance of food safety and quality, there is a need to transform agri-food systems from a capital and commercial industry model to a common-good business model (Wang and Shi 2023). To realize that, besides the establishment of an agroecological compensation mechanism, consumers need to increase their social understanding of agri-food value chains to stimulate authentic support (Bernal 2011; Riches 1999). With such a change in consumer understanding, a new agri-food business model called community supported agriculture has emerged all over the world to call on the closer interaction between the producers (the farmers) and the consumers. This builds mutual trust that was lost due to the spatial and functional separation encouraged by the many middlemen (Shi et al. 2011). The reestablishment of mutual trust between farmers and consumers via community supported agriculture is an excellent agricultural process, without such trust, the CSA agricultural business model will be disingenuous and discarded by the consumers. Therefore, the concerted development of community supported agriculture and eco-agriculture (Wang 2019) or we could call it farmer-community connection agriculture (Brodt et al. 2006) or placed-based agriculture (Swagemakersa et al. 2019; Ikerd 2019; Mullinix et al. 2019) will be the only way to reach the goal of agri-food system transformation in response to climate change, that is why it was named as agroecological transformation or agroecological transition (Schiller et al. 2019).

Similarly with eco-agriculture development, the development process of community supported agriculture in China is worth elaboration. As the second largest population country in the world, China has placed food production as a governance priority in response to the serious questions 'who will feed China' asked by Lester Brown in 1995 (Brown 1995). In this regard, the contradiction between food production and environmental degradation has existed for a long time but has never been more serious than current decades when conventional agriculture practices have increased to produce more food for the increasing population. In that period, food safety problems accumulated unnoticed and finally led to disastrous consequences around the year of 2008 when many children were chronically poisoned by several branded milk powders in which a kind of protein like compound named melamine was added. Such accidents finally awakened the consumers, and in 2009, the first CSA named 'little donkey' was set up in a suburb of Beijing City (Shi et al. 2011), since then, CSA was built nationwide. The products of community supported agriculture are not specifically produced via eco-agriculture and organic farming but are considered 'the hazard free product' at the basic quality line (Scott et al. 2014; Si 2017). The standard of eco-agriculture product is far higher than 'the hazard-free product' which has been abolished, instead, eco-agriculture and its product is being set up as a national standard since 2020 in China (Xu et al. 2024). Our idea is to coordinate the development of eco-agriculture and community supported agriculture to ultimately promote rural economy and increase farmer's income. The multiple goals are: (1) establishing a sustainable agricultural production system via eco-agriculture and strengthening the guiding role of ecological protection; (2) fostering community-supported agriculture to rationalize the agricultural management system, thereby highlighting the principal status of farmers (producers) in agricultural production; and (3) coordinating eco-agriculture and community supported agriculture to achieve the coordinated development of urban and rural areas, embodying the concept of win-win cooperation towards the coming of agroecological transformation (Wang 2019).

In conclusion, when we look back the turning point of agroecology as a practice and as a social movement proposed by Wezel et al. (2009) and reiterated by Olivier de Shutter in 2011, in their idea, agroecology actually is synonymous with eco-agriculture, implying that agroecology as practice or eco-agriculture under the school of agroecology as science can be used to unify all other many alternative agriculture models such as regenerative agriculture, climate-smart agriculture, and conservation agriculture for rebuilding sustainable agriculture system. This is why the authors intended to use this chapter as the terminal chapter of our book. If we are to expect good products from farms, then farmers will have to receive a fair return for those products. Otherwise, no one would like to do eco-agriculture. We cannot wait for sound policy from government (agroecological compensation) and good price from market (eco-labelled product) before we put eco-agriculture into practice. The effective way is educating consumers to recognize and use their power by supporting and financing farmers to do eco-agriculture, in this regard, Community Support Agriculture is the key to make eco-agriculture development flourish worldwide under the umbrella of agroecology.

References

Altieri, M.A. 2004. *Agroecology versus eco-agriculture: balancing food production and biodiversity conservation in the midst of social inequity*. Commission on environmental, economic and social policy occasional papers issue 3. Gland, Switzerland, International Union for the Conservation of Nature.

Bernal, E. 2011. Business innovation and ecological agriculture in Los Monegros (Aragon, Spain) in the 20th century. *Historia Agraria* 53:129–170.

Brodt, Sonja, Gail Feenstra, Robin Kozloff, Karen Klonsky, and Laura Tourte. 2006. Farmer-community connections and the future of ecological agriculture in California. *Agriculture and Human Values* 23:75–88.

Brown, Lester R. 1995. *Who will feed China? Wake up call for a small planet*, World Watch Institute, Environmental Alert Series, 163. New York: W.W. Norton & Company, Inc.

Buck, Louise E., and Jeffrey C. Milder. 2011. Moving eco-agriculture into the mainstream. In *State of the world: innovations that nourish the planet*, ed. L. Starke, 15–24. Washington, DC: The Worldwatch Institute. https://www.researchgate.net/publication/305315832_Moving_Ecoagriculture_into_the_Mainstream. Accessed May 13, 2025.

Buck, Louise E., Thomas A. Gavin, David R. Lee, Norman T. Uphoff, Diji Chandrasekharan Behr, Laurie E. Drinkwater, W. Dean Hively, and Fred R. Werner. 2004. *Eco-agriculture: a review and assessment of its scientific foundations*. Ithaca, NY: Cornell University. api.ecoagriculture.org/uploads/12_de27a4a0f5.pdf. Accessed May 8, 2025.

Buck, Louise E., Jeffrey C. Milder, Thomas A. Gavin, and Ishani Mukherjee. 2006. *Understanding eco-agriculture: a framework for measuring landscape performance*. https://www.researchgate.net/publication/309731592_Scherr_et_al_2014. Accessed May 8, 2025.

Cheatham, Walter Beverly, III. 2022. *Finding Arcadia: ecological agriculture and the art of being in community*. Victoria, BC: Royal Roads University.

Chen, Kai-Lih, Wei-Hsin Kong, Chi-Cheng Chen, and Je-Liang Liou. 2021. Evaluating benefits of eco-agriculture: the cases of farms along Taiwan's east coast in Yilan and Hualien. *Sustainability* 13:10889. https://doi.org/10.3390/su131910889.

Chitakira, M., L. Nhamo, E. Torquebiau, J. Magidi, W. Ferguson, S. Mpandeli, K. Mearns, and T. Mabhaudhi. 2022. Opportunities to improve eco-agriculture through transboundary governance in transfrontier conservation areas. *Diversity* 14:461. https://doi.org/10.3390/d14060461.

Cornelia, Adamov Tabita, T. Iancu, Nagy Andrea, and Toader Cosmina-Simona. 2008. Agricultura Ecologică—Un Model De Agricultură Durabilă (The ecological agriculture—a model of durable agriculture). *Lucrări stiinŃifice Zootehnie si Biotehnologii* 41 (1): 606–611.

de Schutter, Olivier. 2011. *Agroecology and right to food*. Report presented at the 16th Session of the United Nations Human Rights Council[R/OL]. http://www.srfood.org/images/stories/pdf/officialreports/20110308_a-hrc-16-49_agroecology_en.pdf. Accessed June 17, 2024 in Canada.

de Shutter, Olivier. 2010. *Agroecology and right to food*. Report presented at the 16th Session of the United Nations Human Rights Council[R/OL]. Human Rights Council of the United Nation

(2010-12-20) http://www.srfood.org/en/reportagroecology-and-the-right-to-food. Accessed November 20, 2024.

Erlich, P., and A. Erlich. 1981. *Extinction. The causes of the disappearance of species*, 305. New York: Random House.

Fallas Henriquez, A., and S. Molina-Murillo. 2017. Methodological proposal to quantify and to compensate the agroecosystem services generated by the good agricultural practices of small-farmers. *Ecosistemas: Revista Cietifica y Tecnica de Ecologia y Medio Ambiente* 26 (3): 89–102.

Food and Agriculture Organization. 2007. *The state of food and agriculture: paying farmers for environmental services*. FAO Agriculture Series, No. 38, Rome, p 222.

Francis, C., G. Lieblein, S. Gliessman, T. A. Breland, N. Creamer, R. Harwood, and R. Salvador. 2003. Agroecology: the ecology of food systems. *Journal of Sustainable Agriculture* 22 (3): 99–118.

Gliessman, Stephen R. 1998. *Agroecology: ecological processes in sustainable agriculture*. Chelsea, MI: Ann Arbor Press.

Gliessman, Stephen R. 2007. *Agroecology: the ecology of sustainable food systems*, 384. New York: CRC Press, Taylor & Francis.

Ikerd, J. 2019. Place-based food and farming systems: reconnecting people with purpose and place. *Journal of Agriculture, Food Systems, and Community Development* 9 (Suppl. 1): 67–76.

Jitareanu, A. F., M. Mihaila, A. D. Robu, F. D. Lips, and C. L. Costuleanu. 2022. Dynamic of ecological agriculture certification in Romania facing the EU organic action plan. *Sustainability* 14:11105. https://doi.org/10.3390/su141711105.

Kiley-Worthington, M. 1981. Ecological agriculture. What it is and how it works. *Agriculture and Environment* 6:349–381.

Larisa, Cremeneac, and Boclaci Tatiana. 2008. Tehnologia Viermicultivarii—Factor Important In Dezvoltarea Agriculturii Ecologice (The technology of worm culture—the important factor in development of ecological agriculture). *Lucr stiinŃifice Zooteh Biotehnol* 41 (2): 287–290.

Lescourret, Françoise, Thierry Dutoit, Freddy Rey, François Côte, Marjolaine Hamelin, and Eric Lichtfouse. 2015. Agroecological engineering. *Agronomy for Sustainable Development* 35:1191–1198.

Liao, Feng. 2020. Regional public Brands of Agricultural Products in mountain areas from the perspective of value realization of ecological products: a case analysis on Lishui Shangeng. *Journal of Lishui University* 42 (6): 1–10.

Ma, Shijun, and Songhua Li. 1987. *Agroecological engineering in China*, 1–3. Beijing: Science Press.

Ma, Shengna, and Miaoyu Zhuge. 2024. Ecological agriculture traits: constructing an integrated agricultural system for sustainable development. *Geographical Research Bulletin* 3 (16): 2–5.

McNeely, Jeffrey A., and Sara J. Scherr. 2001. *Common ground, common future: how eco-agriculture can help feed the world and save wild biodiversity*. Washington, D.C.: IUCN, Future Harvest.

McNeely, Jeffrey A., and Sara J. Scherr. 2002. *Ecoagriculture: strategies to feed the world and save wild biodiversity*, 323. Washington, DC: Island Press.

Mercier, J. R. 1978. Towards ecological agriculture. *Futuribles* 16:479–484.

Moucheng, Liu, Xiong Yin, Yuan Zheng, Min Qingwen, Sun Yehong, and Anthony M. Fuller. 2014. Standards of ecological compensation for traditional eco-agriculture: taking Rice-fish system in Hani terrace as an example. *Journal of Mountain Science* 11 (4): 1049–1059.

Mullinix, K., N. Robert, and R. Harbut. 2019. Place-based food systems: making the case, making it happen. *Journal of Agriculture, Food Systems, and Community Development* 9 (Suppl. 1): 1–3.

Némethová, Jana. 2010. Ekologické Poľnohospodárstvo V SR (Ecologic Agriculture in Slovakia). *Geografické štúdie* 1:49–60.

Nogrady, Bianca. 2024. Farming techniques that protect the planet. *Nature* 630:S23–S24.

Patarlageanu, Simona Roxana, Mihai Dinu, Amelia Diaconu Artifex, and Mihaela Diana Oancea Negescu. 2022. Ecological agriculture and its role in sustainable development. *Proceedings of the International Conference on Business Excellence* 16 (1): 390–399.

Regnier-Davies, J. 2015. *'Fake meat and cabbageworms': connecting perceptions of food safety and household level food security in urban China.* Master's thesis, University of Waterloo.

Riches, Graham. 1999. Reaffirming the right to food in Canada: the role of community-based food security. In *For hunger proof cities. Sustainable urban food systems*, ed. M. Koc, R. McRae, and J. Welsh, 203–207. Ottawa: IDRC Press.

Sanders, Richard. 2000. Political economy of Chinese ecological agriculture: a case study of seven Chinese eco-villages. *Journal of Contemporary China* 9 (25): 349–372.

Sandu, Mariana, Amalia Strateanu, and Lavinia Udrea. 2023. Analysis of ecological agriculture from the perspective of maintaining the biodiversity of agricultural lands. *Annals "Valahia" University of Targoviste – Agriculture* 15 (1): 22–26.

Scherr, Sara J., and Jeffrey A. McNeely, eds. 2007. *Farming with nature. The science and practice of eco-agriculture*, 445. Washington, DC: Island Press.

Scherr, Sara J., and Jeffrey A. McNeely. 2008. Biodiversity conservation and agricultural sustainability: towards a new paradigm of 'eco-agriculture' landscapes. *Philosophical Transactions of the Royal Society of London. Series B, Biological Sciences* 363:477–494.

Scherr, S. J., L. Buck, L. Willemen, and J. C. Milder. 2014. Eco-agriculture: integrated landscape Management for People, food, and nature. In *Encyclopedia of Agriculture and Food Systems*, ed. Neal Van Alfen, vol. 3, 1–17. Amsterdam: Elsevier.

Schiller, Katharina J. F., Laurens Klerkx, P. Marijn Poortvliet, and Wendy Godek. 2019. Exploring barriers to the agroecological transition in Nicaragua: a technological innovation systems approach. *Agroecology and Sustainable Food Systems* 44 (1, 2): 88–132. https://doi.org/10.1080/21683565.2019.1602097.

Scholes, Mary. 2012. After Durban: new issues on the table. *Environment and Development* 1:124–125.

Scott, Steffanie, Zhenzhong Si, Theresa Schumilas, and Aijuan Chen. 2014. Contradictions in state- and civil society-driven developments in China's ecological agriculture sector. *Food Policy* 45:158–166.

Shi, Tian. 2003. Moving towards sustainable development: rhetoric, policy and reality of ecological agriculture in China. *International Journal of Sustainable Development & World Ecology* 10 (3): 195–210.

Shi, Tian, and Roderic Gill. 2005. Developing effective policies for the sustainable development of ecological agriculture in China: the case study of Jinshan County with a systems dynamics model. *Ecological Economics* 53:223–246.

Shi, Yan, Cunwang Cheng, Peng Lei, Tiejun Wen, and Caroline Merrifield. 2011. Safe food, green food, good food: Chinese community supported agriculture and the rising middle class. *International Journal of Agricultural Sustainability* 9 (4): 551–558.

Si, Zhenzhong. 2017. Rebuilding consumers' Trust in Food: community-supported agriculture in China. In *Sustainable food futures: multidisciplinary solutions*, ed. J. Duncan and M. Bailey, 34–45. London: Routledge.

Song, Chengjun, Renhua Sun, Zuliang Shi, Yinghao Xue, Jiuchen Wang, Zhiyu Xu, and Shangbin Gao. 2022. Construction process and development trend of ecological agriculture in China. *Acta Ecologica Sinica* 42:624–632.

Sriskandarajah, Nadarajah, Charles Francis, Lennart Salomonsson, Helena Kahiluoto, Geir Lieblein, Tor-Arvid Breland, Ulrika Geber, and Juha Helenius. 2006. Education and training in ecological agriculture: the Nordic region and the USA. In *Organic agriculture—a global perspective*, ed. Lennart Salomonsson and Ulrika Geber. Clayton, NC: CSIRO Publishing. https://res.slu.se/id/publ/9701.

Sun, Hongliang, and Ye Qi. 2017. From eco-agriculture to eco-civilization construction: in memory of prof. Ma Shijun's 100 years birthday and the 36th anniversary of eco-engineering theory publication. *Chinese Journal of Eco-Agriculture* 25 (1): 8–12.

Swagemakersa, Paul, M. Dolores Domínguez Garcíaa, Pierluigi Milone, Flaminia Ventura, and Johannes S. C. Wiskerke. 2019. Exploring cooperative place-based approaches to restorative agriculture. *Journal of Rural Studies* 68:191–199.

Teodorescu, Gabriela. 2008. Agricultura Ecologică—Noi Perspective In Romania (Ecological agriculture—new perspective for Romania). *Lucr stiinŃifice Zooteh Biotehnol* 41 (1): 256–260.

Tomich, Thomas P., Sonja Brodt, Howard Ferris, Ryan Galt, William R. Horwath, Ermias Kebreab, Johan H. J. Leveau, Daniel Liptzin, Mark Lubell, Pierre Merel, Richard Michelmore, Todd Rosenstock, Kate Scow, Johan Six, Neal Williams, and Louie Yang. 2011. Agroecology: a review from a global-change perspective. *Annual Review of Environment and Resources* 36:193–222.

Ursu, A., and I. L. Petre. 2022. Forecasting the optimal sustainable development of the Romanian ecological agriculture. *Sustainability* 14:14192. https://doi.org/10.3390/su142114192.

Vermeulen, Sonja J., Bruce M. Campbell, and John S. I. Ingram. 2012. Climate change and food systems. *Annual Review of Environment and Resources* 37:195–222.

Wang, S. L. 2005. Information technology: toward the way to the sustainable management of agroecosystem. *Agriculture Network Information* 8 (4–7): 12.

Wang, S. L. 2019. Concerted development of ecological agriculture along with community-supported agriculture to facilitate rural vitalization. *Chinese Journal of Eco-Agriculture* 27 (2): 212–217.

Wang, S. L., and D. M. Chen. 2009. Development of modern ecological agriculture in Fujian: achievements, problems and countermeasures. *Journal of Fujian Agriculture and Forestry University* 12 (4): 25–29.

Wang, S. L., and W. X. Lin. 1999. Ecological agriculture in China and sustainable agriculture in the world have the same goal. *Research of Agricultural Modernization* 20 (2): 81–84.

Wang, S. L., and S. X. Shi. 2023. Developing Chinese ecological agriculture acts as a vital pathway of the state-specifically agricultural modernization in China: a historical review of ecological agriculture movement in China. *Chinese Journal of Eco-Agriculture* 31 (8): 1184–1193.

Wang, S. L., C. D. Caldwell, and W. F. Zhu. 2010. Low carbon agriculture: origins, principles and strategies. *Research of Agricultural Modernization* 31 (5): 604–607.

Wang, S. L., C. D. Caldwell, S. L. Kilyanek, and Sean M. Smukler. 2019. Using agroecology to stimulate the greening of agriculture in China: a reflection on 15 years of teaching and curriculum development. *International Journal of Agricultural Sustainability* 17 (4): 298–311.

Wezel, A., S. Bellon, T. Doré, C. Francis, D. Vallod, and C. David. 2009. Agroecology as a science, a movement and a practice. A Review. *Agronomy for Sustainable Development* 29 (4): 503–515.

Xu, Xiangbo, Xu Rong, Zhang Linxiu, Hu Xiaofang, Li Xiaoyang, Xue Yinghao, and Xu Zhiying. 2024. Establishment of ecological farms in the development of ecological agriculture: historical perspective, current progress, and future outlook (in Chinese). *Chinese Journal of Eco-Agriculture* 32 (4): 701–712.

Ye, Qianji. 1982. Ecological agriculture. *Problems of Agricultural Economy* 11:3–10.

Ye, Qianji. 1988. *Ecological agriculture: the future of agriculture*. Chongqing: Chongqing Publishing Group.

Ye, Qianji, and Biliang Luo. 1987. A preliminary study on the strategy of the development of eco-agriculture in China. *Journal of Southwest Agricultural University* 9 (1): 1–8.

Zhang, Qian Forrest. 2024. From sustainable agriculture to sustainable Agrifood systems: a comparative review of alternative models. *Sustainability* 16:9675.

Zhu, Wenfeng. 2011. *Study on assessment of Agroecosystem Service & Health and their managerial model in Fujian Province*. Fujian Agriculture and Forestry University.

Epilogue: The 'Motivating Spring' of 2004

Better is only possible if you believe it's possible and fight for it.

—Alex Himelfarb, May 16, 2025

After almost 2 years of devotion to this *Advanced Agroecology* textbook since the climax of pandemic COVID-19 in early 2023 in China, I am finally released with this manuscript being submitted. However, the moment makes me flashback to the first meeting with Dr. Claude Caldwell at the spring of 2004 (Photo 1), which I like to call a 'Motivating Spring'. Since then, we have been engaging in our collaborative teaching of agroecology for 21 years until we had the last lesson on April 30th of 2024 (Photo 2). Here is the story and its background.

Just prior to our meeting, the two authors' affiliations, respectively, Fujian Agriculture and Forestry University (FAFU) in China and Nova Scotia Agricultural College (NSAC, merged to Dalhousie University as the latter's Faculty of Agriculture in 2012) in Canada, had established an effective cooperative education platform in the field of agriculture with official approval since 2003. One of the authors of this book, Dr. Claude D. Caldwell, as the Canadian side founder of the program, initially introduced the agroecology course from NSAC to the first class of 2+2 programmed students enrolled at FAFU. It was originally aimed at helping students to become familiar with and adapt to Canadian universities' teaching style and pedology. I (Songliang Wang) fortunately was selected by the FAFU side as a agroecology course teaching partner. In our first agroecology class in February 2004 in a FAFU classroom, I was shocked when Dr. Claude Caldwell told the students his new definition of agriculture shown on a PowerPoint slide for 'Unit 1: Introduction to Agroecology Course':

> Agriculture is the science, art, politics, and sociology of changing sunlight into healthy and happy people.

Yes, why not! Such new understanding tells why and how we humans conduct the agriculture, why? for the health and happiness of people and how? by means of

S. Wang, C. D. Caldwell, *Advanced Agroecology*,
https://doi.org/10.1007/978-3-032-08689-1

Photo 1 The two authors of this book first met in FAFU classroom in February 2004

Photo 2 The two authors of this book had their last on-site class in FAFU classroom on April 30, 2024

exploring plants to convert sunlight into food. By using this lens to distinctly reflect on the current capitalized and industrialized agriculture, it becomes obvious that agriculture is moving in the wrong direction. At the moment of hearing the new definition of agriculture, I suddenly realized that it is a concept that is worthy for everyone to know and to understand. Accepting this concept and learning to practice it might just change the world in agriculture and the affairs of food. It was reminiscent for me of the first time I read Rachel Carson's Silent spring which produced a global awakening of people for the environment. It was then that I made up my mind to let more people know this new understanding or definition of agriculture. It was a motivating spring, at least for me and now I conclude by composing the current textbook that it has been motivating me for co-teaching agroecology with Claude and Shannon Kilyanek for 21 years.

The authors of this textbook individually started to teach agroecology earlier than this new definition of agriculture was conceived. Claude began his teaching agroecology career in 1980s, and Songliang started his in the 1990s, but only in that moment did I realize that agroecology will be a discipline and science that brings a paradigm change to reach the ideal goal of agriculture. Coincidentally and fortunately, in the USA and Nordic Europe, a group of agroecologists might be bearing the same idea; they were at the same time scaling out agroecology from a marriage of ecology to agronomy, ecology for alternative agriculture, ecological basis of sustainable agriculture, to the ecology of entire food system (Francis et al. 2003). A little bit later in the 2000s, a group of agroecologists in France were scaling up agroecology as a science, a practice, and a social movement (Wezel et al. 2009). As is mentioned in this text, the definition and application of agroecology have subsequently evolved into a theory, practice, and social movement.

Meanwhile, our small international agroecology teaching group also scaled up and out, firstly, as our co-teaching collaboration evolved, more and more outstanding talented students were educated with both the idea and approaches of agroecology. The positive impact of these talented students continues in both China and Canada and around the world. Truly education is the key to change. Achievement of our teaching was summarized and published and we have been expanding our collaboration from teaching to research and application, study on agroecosystem service assessment, recommending policy implications and practice in eco-agriculture. We have co-published ten academic papers and articles in the field of agroecology, and three textbooks in agroecology:

- Songliang Wang, Claude D Caldwell, Shannon L. Kilyanek & Sean M. Smukler. Using agroecology to stimulate the greening of agriculture in China: a reflection on 15 years of teaching and curriculum development. International Journal of Agricultural Sustainability, 2019,17(4):298–311
- Songliang Wang, Claude D Caldwell, Wenfeng Zhu. Low carbon agriculture: origins, principles and strategies (in Chinese). Research of Agricultural Modernization, 2010,31(5):604–607
- Songliang Wang, Claude D Caldwell, Ecology: A core science in Eco-civilization era— exemplified with the roles of rural ecology in guiding the beautiful village

construction (in Chinese). Journal of Fujian Agriculture and Forestry University，2016,19(3): 14–18

- Songliang Wang, Claude D Caldwell, Shannon L. Kilyanek, Dexi Lin, Liqing Wei. Investigation and practice of the elite, bilingual course of agroecology established under the Sino-Canadian collaboration (in Chinese). Journal of Fujian Agriculture and Forestry University，2010,13(1):90–95
- Songliang Wang, Claude D Caldwell, Liqing Wei, Haiyan Su. Ten Years' Chinese-Canadian Collaboration in Undergraduate Education in Fujian Agriculture and Forestry University (FAFU) of China: Curriculum Development. Frontiers of Education in China, 2015,10(3):427–438
- Songliang Wang and Claude D Caldwell. Application of ecological thinking in reconstructing traditional agronomy: A case study of agroecosystem management as core applied subject (in Chinese). Chinese Journal of Ecological Agriculture, 2013,20:39–46.
- Wenfeng Zhu, Songliang Wang and Claude D Caldwell. Agroecosystem service and its managerial essence (in Chinese). Chinese Journal of Eco-Agriculture, 2010,18(4):889–896.
- Wenfeng Zhu, Claude D Caldwell, Songliang Wang et al. A Sino-Canadian Comparative Study of Experts' Perceptions towards Agroecosystem Health. Human Ecology, 2021,49:81–89.
- Wenfeng Zhu, Songliang Wang, Claude D Caldwell. Pathways of assessing agro-ecosystem health and agroecosystem management. Acta Ecologica Sinica, 2012,32:9–17.
- Zhiming Kang，Bin Wang，Wenfeng Zhu，Songliang Wang , Claude D Caldwell and Jianming Wang. Spatial Distribution of Heavy Metals in Agricultural Land Soil and Evaluation of Potential Ecological Risk in Mingxi County, Fujian Province (in Chinese). Journal of Fujian Agriculture and Forestry University，2015,44(2):212–218.
- Claude D Caldwell, Songliang Wang(Co-editors). Introduction to Agroecology. Springer, 2020-12-1，ISBN:9787030684158.
- Songliang Wang, Claude D Caldwell, et al. Agroecology. China Science Press, 2012-8-1, ISBN: 9787030329998.
- Songliang Wang, Claude D Caldwell, et al. Agroecology(the second edition). China Science Press, 2021-6-1, ISBN: 9789811588358.

Among them, the textbook entitled *Introduction to Agroecology* was published by Springer in December 2020, which accommodates the new context of the agro-ecology, opening doors to undergraduates and fresh readers of agroecology. As a successor to the abovementioned textbook for undergraduates, herein we are submitting a textbook entitled *Advanced Agroecology*; the targeted audience for this textbook will be postgraduates and their instructors in agriculture-associated universities and colleges worldwide. The goal of this textbook was to clearly deliver to the readers and users: (1) the problems of our agriculture and food system globally and regionally using agroecology as the lens; (2) processes and techniques for understanding and re-evaluating rural landscapes (ecosystems) and farming systems; and (3) the pedagogy of agroecology teaching and methodology for agroecology studies, using experiential learning and action research of agroecology at the graduate education level by the authors and from worldwide. It is worth to mention that both collaborating bodies, FAFU and Dalhousie University, have promoted the platform into postgraduate-level education collaboration, accommodated by the Internationally Joint Dalhousie College in FAFU during we are composing this textbook, which will become the practical testimony and authentication of this new textbook for the programmed postgraduates.

I hope this book stimulates the truly and lively thinking and discussion for our future agri-food system and correspondingly educational and paedological transitions in world's universities and colleges.

Dr. Songliang Wang
(June 3rd, 2025, in author's office of FAFU)

References

Francis C., Lieblein G., Gliessman S., Breland T., Creamer N., Harwood R., Salomonsson L., Helenius J., Rickerl D., Salvador R., Wiedenhoeft M., Simmons S., Allen P., Altieri M., Flora C., Poincelot R.2003.Agroecology: The ecology of food systems. Journal of Sustainable Agriculture 22, 99–118.

Wezel, A., Bellon, S., Doré, T., Francis, C., Vallod, D., & David, C. 2009. Agroecology as a science, a movement and a practice[J]. A Review. Agronomy for Sustainable Development, 29(4), 503–515.

Index

S. Wang, C. D. Caldwell, *Advanced Agroecology*,
https://doi.org/10.1007/978-3-032-08689-1

MIX
Papier aus verantwortungsvollen Quellen
Paper from responsible sources
FSC® C105338

If you have any concerns about our products,
you can contact us on
ProductSafety@springernature.com

In case Publisher is established outside the EU,
the EU authorized representative is:
Springer Nature Customer Service Center GmbH
Europaplatz 3, 69115 Heidelberg, Germany

Printed by Libri Plureos GmbH
in Hamburg, Germany